HISTORY OF FREE FALL

Aristotle to Galileo

With an Epilogue on π in the Sky

To my friends of more than three decades
A. Rupert Hall and Marie Boas Hall
this monograph is dedicated
with affection and respect

HISTORY OF FREE FALL
Aristotle to Galileo
With an Epilogue on π in the Sky

Stillman Drake
University of Toronto

WALL & THOMPSON
Toronto

Chapter 5, "Discovery of the Law of Fall," is essentially duplicated in the Introduction to *Two New Sciences*, 2nd edn. (Wall & Thompson, 1989) and in *Galileo: Pioneer Scientist* (in Press).

Canadian Cataloguing in Publication Data
Drake, Stillman
History of free fall

Bibliography: p.
ISBN 0-921332-26-2

1. Motion – History – Sources.
2. Mathematical physics – History.
3. Physics – History – 17th century.
4. Galilei, Galileo, 1564-1642.
I. Title.

QC123.D73 1989 531'.5'09 C89-094555-1

IDEA FOR COVER DESIGN: Lorne Perry, Illustrator, *Alberta Science Centre.*

ISBN 0–921332–26–2
Printed in Canada.
1 2 3 4 5 93 92 91 90 89

Table of Contents

1	Introductory	1
2	From Aristotle to Hipparchus, 350 –150 B.C.	7
3	From 1300 to 1500	13
4	The Sixteenth Century	23
5	Discovery of the Law of Fall	35
6	Applications of the Law	51
7	Reception of the Law, 1632–49	67
	Epilogue: Kepler's Problem and Galilean Units	81
	Bibliography of Works Cited	93
	Index	97

1
Introductory

The law of fall states that distances from rest are as the squares of the elapsed times. Galileo used this in his *Dialogue* of 1632 and developed consequences of it in *Two New Sciences* six years later. The law had been first recognized by him early in 1604, immediately following his discovery of the pendulum law.

In a way it is surprising that the times-squared law for the spontaneous descent of heavy bodies had not been recognized long before the 17th century. Measurements sufficient to put the law within someone's grasp are quite simple.[1] Equipment for making them had not been lacking—a gently sloping ramp, a heavy ball, and the sense of rhythm with which everyone is born. Discovery of the law by those means would be easier from rough measurements than from painstaking ones. Counting regularly would suffice for timing. The ball's place at each count is marked, and a piece of string, stretched between the thumbnails from the point of rest to the first mark, is used as the unit of length. The marks will thus be found to be separated, serially, by 1, 3, 5, 7, ... units. Distances from rest must then sum to 1, 4, 9, 16, ...; and, since the counts 1, 2, 3, 4, ... numbered equal times, the times-squared rule of distances from rest becomes immediately recognizable.

If one were to measure in fractions of an inch, applying a yardstick instead of a length of string, this odd-number spacing might go unnoticed and the law would not become evident.[2] The unit used makes no

---

[1] The matter of proof, or rigorous derivation of the law, is quite separate from that of its discovery, or recognition, from measurements and calculations.

[2] The odd-number rule will be found unmistakably when the first distance is used as the unit, even in casual demonstrations on a gently sloping plane, as by holding the hand stretched to the first distance and counting its applications to the others.

difference mathematically, but could greatly increase the difficulty of discovery. It happens that the best unit in this case is the natural one for a person merely curious, not seeking a cause. Idle curiosity might thus have led to the discovery of the law of fall at any epoch (and more probably by a construction foreman than by a natural philosopher.)

Once the times-squared law has been so found, a conjecture that the same rule applies also to vertical fall could be easily verified, to any practicable degree of accuracy. On the other hand it would be extremely difficult for anyone to discover the fact in the same way that it could easily be shown to be true. Furthermore, however nearly the law might be found to correspond with actual measurements, it is quite another matter to derive the times-squared law by logic, or to prove it mathematically.

The ease of stumbling on this discovery renders it highly improbable that natural philosophers had ever searched for the *law* of fall. They gave many reasoned accounts of the *cause* of fall, from Aristotle's to Buridan's, all without value in terms of useful knowledge.[3] Newton lamented his inability to find the cause of gravitation. It may be hard to imagine what *kind* of a cause he was looking for, and it has been a waste of time for historians to seek a historical source of the *law* of fall among philosophical quests for the *cause* of fall that were made up to the time of Galileo—and also by him, in his early years.

Historians were in possession of the law of fall when they began the search for its possible sources before the 17th century. Knowing proportionality to be involved in it, it seemed to them reasonable to assume that some natural philosopher before the time of Galileo had hit upon the idea that a proportionality of speeds in accelerated motion to distances fallen (or to times of fall) could explain the motions of heavy bodies causally. But that was simply not the case. Aristotle had defined the goal of science to be causal knowledge. Any proportionality capable of accounting causally for fall would, it seemed to his followers,

---

[3] Galileo's comment on the subject in his famous *Dialogue* is worth quoting. His spokesman having said that he would know what makes the earth move if he knew what moves heavy bodies downward, the spokesman for philosophers replied that that cause is well known; it is gravity. "You are wrong," was the answer; "what you ought to say is that it is *called* gravity. I am asking not for its name, but its essence, of which you know no more than you know the essence of whatever moves the planets."

necessarily relate speeds either to the heaviness of a falling body or to some external force impressed upon it. Those things were acceptable causal agencies, whereas the distances or times of a motion were philosophically unacceptable as its cause. Heaviness as the cause of fall, and impressed force as the cause of acceleration during fall, were proposed (in that order) by theories of fall during Greek antiquity. The two alternative ancient causes were eventually combined in a novel 14th-century theory of fall. The 16th century witnessed the destruction of simple heaviness as the cause of fall, by mathematical logic and by actual tests. The heavier of two bodies does not fall with a speed proportionately greater than speed of the lighter. As the 16th century ended, Galileo saw that it was superfluous to assume action of impressed force in order to explain continuation of any motion. Only then, with those traditional causal preconceptions out of the way, was the times-squared law—which might easily have been found through curiosity—discovered by Galileo after a series of painstaking and rather complex investigations. The discovery followed immediately upon his discovery of the law of the pendulum, which in turn had required his adoption of units of length and time particularly appropriate for gravitational phenomena. Although Galileo did not apply them to planetary motions, that will be done in the Epilogue on Kepler's problem, for the Galilean units are still of use today.

Before the 17th century was over, Newton restored both the old ideas of heaviness and impressed force at a higher level of physical thought. But his creation of modern dynamics came after the end of the period to be considered in the present monograph.

Despite its rigorous derivation in 1638 (and confirmation by measurement), the law of fall was still in dispute among scholars in 1650. The notion that it produced a sudden shift of attention among 17th-century natural philosophers, away from causal inquiry and toward measurements and mathematics, is more a product of our customary terminology than a demonstrable truth. We call René Descartes a natural philosopher; yet far from his having sought consequences of the times-squared law, he declared that Galileo had merely asserted his law without any causal foundation, and that it rarely, if ever, applied in actual fall. Honoré Fabri, who vigorously opposed the new law of fall, is called a natural philosopher—but so are Christiaan Huygens and Isaac Newton who as vigorously investigated its consequences, strengthening its foundations as a true law of nature.

Publication of the times-squared law of fall did not hinder philosophical speculations about the cause of spontaneous motion by heavy bodies, or improve them. In physics, it initiated an unprecedented line of inquiry into natural phenomena. Confusion arises from calling those who adhered to strictly causal theories of fall by the old, established name, and also extending the same title to those others who investigated physical phenomena in the new way, aimed toward a different goal. This confusion could be eliminated by calling the former "natural philosophers" and the latter "physicists," or "scientists," or the like. Yet Newton certainly, and Huygens very probably, thought of himself as a natural philosopher, so we do well not to alter our terminology. Any confusion is better avoided by understanding clearly what *did* happen in the 17th century—no matter what status that may have had then (or now has) among philosophers of the highest repute.

The principal purpose of the present monograph is to improve our understanding of what did happen in the 17th century. To do that effectively it is necessary to begin at the beginning, from theories of fall and rules arising incidentally to them—rules which, being wrong, are not entitled to be called "laws" without mental reservations. When the old theories and rules are studied in order, they show a certain pattern of recurrence at each epoch in the history of theories of fall. In turn, some events of the 17th century then appear less surprising, and less revolutionary in the usual sense, than they now seem to have been, by becoming more easily understandable in this chronological framework.

Two obstacles exist to comprehension of that framework. One is the fact that theories of fall from Aristotle to Galileo have been fully documented and studied by many scholars. The prospect of learning anything new seems hardly to justify one's reviewing the old documents and reconsidering the established conclusions. Nevertheless, patient re-examination of familiar things has often resulted in historical insights, formerly unappreciated only by reason of some inveterate preconception.

The other obstacle is that events of the 17th century have been even more copiously documented than those of ancient and medieval natural philosophy, so that nothing relevant to the law of fall seems to have escaped notice. Nevertheless, a decade ago there was still uncertainty about the circumstances of the eventual discovery. From Galileo's working papers, restored to their order of composition, it is at last possible to follow his work step by step, and hence to determine whether, and how, any of the known theories of fall could have helped or hindered that eventual discovery. That process brings to attention aspects of older theories that had been relatively obscured from notice.

Until quite recently it has been possible only to speculate about the possible influence of this or that theory on Galileo's thought at one time or another during his long career. A great many such speculations were offered, of which none required any role for the pendulum law in the discovery (because pendulums had played no part in old theories of fall.) Now, however, Galileo's working papers leave no doubt that his 1604 discovery of the law of fall depended on his having found the pendulum law. Not only was that used in Galileo's approach, but when G.B. Baliani also published the same law in the same year (1638), he postulated the pendulum law for his proof of the times-squared law of fall. It is clear that not everything that happened in the 17th century to bring forth the times-squared law was already at least implicit in some previous theory of fall. Documents formerly unexamined reveal what had been lacking, after which, with knowledge of the facts in need of explanation, new study of familiar documents is called for to reveal why that had been lacking.

There is also at least one positive reason for composing a monograph on theories of fall; namely, that none already exists. In order to learn the pre-history of the times-squared law of fall it is presently necessary to study the scattered accounts—sometimes contradictory accounts, and so far as the 16th century is concerned, incomplete accounts. Nearly all the documentation needed for theories of fall before 1500 was compiled a quarter-century ago by Marshall Clagett, who incorporated it with a much greater bulk of documentation for many other aspects of natural philosophy and mechanics during the same period.[4] To extract the part relevant to the present topic, and present it together with documentation for the 16th and 17th centuries, will therefore be of some service to historians of early modern physics.

The times-squared law of fall belongs to purely mathematical physics, a field opened by Archimedes in antiquity, approached in a different way during the 14th century, and finally revised and continuously pursued from the 17th century to the present. Old theories of fall belonged to speculative physics, and came only incidentally to include some mathematical formulations.

Clearly, the mathematical physics of any epoch is shaped by the

---

4 Marshall Clagett, *The Science of Mechanics in the Middle Ages* (Madison, 1959.) Citations of documents from this source will be identified by the name Clagett followed by page numbers. Some of his English translations will occasionally be slightly altered in the present text, since comparison with the original Latin in that volume may be easily made.

mathematics of that period. Because there was a fundamental difference between the mathematics employed by Archimedes and that created in the 14th century, while that of Archimedes was reinstated in the 16th century, it will be necessary to identify the nature and occasions of such alterations of mathematics. The relevance of each change to theories (and to the law) of fall may not be immediately clear at the places of exposition, but will become so as the story proceeds.

The language in which old theories of fall were expressed also underwent changes with time, and again in translation from Greek, the language of the first theories, into Latin. Thus the word κίνησις became *motus*, so in modern-language translations of Aristotle, commentaries on his writings, and discussions of his meaning, κίνησις appears as "motion" (or its equivalent in other languages). Aristotle's term included not only locomotion, or change of place with time, but also alteration and growth. The word "physics" was coined by Aristotle to name the science of nature, from φύσις, which became *natura* in Latin, so that the word "natural" (as in natural philosopher) is often best thought of as meaning "physical" in modern usage. It is less simple to deal with linguistic changes (semantic shifts) than with changes which have taken place in mathematics, but they will at least be pointed out as occasion arises.

2

From Aristotle to Hipparchus, 350 –150 B.C.

The law of fall was first stated as proportionality between distances and squares of times, and then derived from a direct proportionality of speeds to times. Whether Aristotle intended any proportionalities in his rules of fall is debatable, but if he did, one was a proportionality of speeds to *weights* when bodies fall through the same medium. A given body falling in different media was (proportionally?) swifter in the less dense medium. Such, at any rate, were the rules taught as Aristotle's from his scattered statements, mainly in *De cælo*. Assuming them to have represented his thought, they did not reveal the hidden cause of acceleration during fall unless the weight of a body increased during the process of falling, and that was the cause offered by Aristotle for greater speed at the end of fall than in its middle part.

Aristotle's theory of fall, though presented piecemeal, was complete and logically coherent. It explained the direction[5] of motion, rules of swiftness, and the cause of acceleration. Equal speed and greater speed were defined terms, implying the meaning of lesser speed. Only swiftness itself remained undefined, as a quality of motion known intuitively. His was a scientific theory in Aristotle's sense, by virtue of its giving a causal account of the fall of heavy bodies.

Aristotle did not invoke force to explain continued motion by a body thrown after it left the hand. He did maintain that anything in motion is moved by something; that is, not by itself, but by contact with its mover. Because only the surrounding air remained in contact with a projectile, he favored the opinion that when moving the thing thrown, the moving hand imparts some of its moving power to the air,

---

[5] Straight toward the center of the universe, at which Aristotle held the earth to also be centered, not by necessity but in fact.

which power continued the motion begun by the hand. Forced motion, being against nature by definition, played little part in Aristotle's physics as the science of nature; projectiles were mentioned only in passing and near the end of his *Physics*.

The first rival theory of fall deserving consideration was proposed about two centuries later by Hipparchus, who was not a philosopher but an astronomer and a mathematician. His book has not survived, but we have its title and a summary of the theory of fall it contained, given by Simplicius in his sixth-century commentaries on Aristotle's *De cælo* as follows:

> Hipparchus in his work entitled *On bodies carried down by heaviness* declares that, in the case of earth thrown upward, the projecting force is the cause of the upward motion as long as the projecting force overpowers the downward tendency of the projectile, and that to the extent that this projecting force predominates, the object moves the more swiftly upward. Then, as this force diminishes, (1) the upward motion continues but no longer at the same rate; (2) the body then moves downward under the influence of its own internal impulse, even though the original projecting force lingers in some measure; and (3) as this force continues to diminish, the object moves downward always more swiftly, and (4) most swiftly when this force is entirely lost.
>
> Now, Hipparchus asserts that the same cause operates in the case of bodies let fall from on high. For, he says, the force which held them back remains with them up to a certain point, and this is the restraining factor which accounts for the slower movement at the start of the fall.[6]

That is another coherent theory of fall, quite different in its implications from Aristotle's, and it appears to be the only other coherent theory put forth in Greek antiquity. Inasmuch as Hipparchus was a mathematician, whereas Simplicius was concerned chiefly with strictly philosophical issues, it is probable that the book contained arguments in support of this theory of fall in relation to a supposed impressed force of projection that were not included by Simplicius, understandably if they were mathematical in character.

The clause, "...to the extent that this projecting force predominates, the object moves the more swiftly upward," certainly suggests that a proportionality was intended by the mathematician whose theory was being outlined. In upward motion, the degree to which the force ex-

---

[6] Clagett, p. 543.

ceeded the natural tendency downward was always diminishing, until it vanished entirely, after which event the downward motion must become uniform at a speed determined only by Aristotle's cause of fall, the heaviness of the body. Hence, in the new theory, acceleration during fall was of limited duration.

It is natural to wonder why Hipparchus chose not to assume that acceleration would continue throughout fall. Presumably he had reason to regard something as left unexplained, or even as contradicted, by Aristotle's account of fall. One possibility is that, like Philoponus centuries later, Hipparchus knew that in falling through considerable distances, different weights strike the ground nearly together.[7] What Philoponus wrote in his sixth-century commentaries on Aristotle's *Physics* was:[8]

> If you let fall from the same height two weights, of which one is many times as heavy as the other, you will see that the ratio of the times required for the motion does not depend on the ratio of the weights, but that the difference in time is a very small one.

That would certainly be the case if in fact the bodies fell with nearly equal speeds after the leftover force of projection had vanished. Likewise, observation may account for Hipparchus' connection of the case of fall with that of continued motion in projectiles, as Aristotle had not done. Now, although Philoponus did not make that connection explicitly, in his later discussion of projectile motion in the same commentary, having criticized the explanation favored by Aristotle, Philoponus went on to say:[9]

> ...It is necessary to assume that some incorporeal motive force is imparted by the projector to the projectile, and that the air contributes either nothing at all or else very little to this motion of the projectile. If, then, forced motion is produced as I have suggested, it is quite evident that if one imparts motion contrary to nature, or forced motion, to an arrow or a stone...there will be no need of any [material] agency external to the

---

7 When Galileo was a student at Pisa he disputed the rule taught to him as Aristotle's, because he had seen large and small hailstones striking the ground together. Observations of the kind had always been possible, and doubtless many had been made in antiquity.

8 Clagett, p. 546.

9 Clagett, pp. 509–10.

projector...

The striking similarity between the thought of Hipparchus and that of a commentator on Aristotle centuries later might be regarded by many historians of science as proof that Philoponus had read Hipparchus. On the other hand it is quite possible that a simple observation of falling bodies led each, in turn, to very similar conclusions, contrary to those presented by Aristotle and accepted as scientific explanation by philosophers who cared only for reasoning and neglected to make observations, or considered them to be misleading when they contradicted, or appeared to contradict, pure reasoning from metaphysical principles.

Returning now to Hipparchus, the first author known to have associated the projecting force[10] with the body, we may call his theory of fall the first dynamic explanation of that phenomenon. Aristotle's theory may be called kinetic, for he regarded weight not as a force but as a natural tendency, a realization of the energy of matter. Or it could be called kinematic, as he wrote in terms of motions rather than of forces.

The idea of "force" came into Greek philosophy as a concept of something altering the course of nature, and it was associated with the notion of the supernatural, as when the gods intervene to compel an action (serving a purpose of their own) evidently contrary to the interests of the doer. It was also a principle of Aristotle's that nothing forced, or violent, could long endure. The concept of power was different; powers could exist in nature and need not, like force, be contrary to nature and accordingly excluded from physics (as the science of nature.)

Here we may properly digress to consider a mathematical rule proposed by Aristotle for forced motions. In his *Metaphysics*, Aristotle appeared to exclude the method of mathematicians from physics when he wrote:

> The minute accuracy of mathematics is not to be demanded in all cases, but only in the case of things which have no matter. Hence its method is not that of physical [natural] science, for presumably the whole of nature has matter. (Book α, at end.)

---

[10] What word Hipparchus used is not known; the Latin translation of the passage in Simplicius gives *virtus*, usually rendered as "strength" rather than as force (= *vis*), and closer to "power."

It is not certain from this that Aristotle meant to exclude the method of mathematicians entirely from physics as the science of nature. On the contrary, it is probable that he meant only to caution others against the extreme mathematicism of the "Italian philosophers" (Pythagoreans) and of his teacher Plato, in which geometric forms and certain properties of particular numbers were invoked to explain qualities and to serve as causes. The phrase "only in cases of things that have no matter" suggests the lines, surfaces, and points of mathematicians. But other things might also have been meant, for Aristotle himself made use of a purely mathematical reason for one rule given in his *Physics*:

> If the movent A have moved B a distance Γ in a time Δ, then in the same time the same power[11] A will move ½ B twice the distance Γ, and in ½ Δ it will move ½ B the whole distance Γ; for thus the rules of proportion will be observed[12]...

The continuation (postponed to Chapter 3) gave rise to an important medieval development in mathematics. Here the point of interest is that the rule just given was for locomotion, to which Aristotle gave priority over other forms of κίνησις, or change, that he placed at the heart of physical science. Even Aristotle himself had not flinched from a rule of locomotion grounded only in the method of mathematics.

Now, if one stops to think of this, it is not unreasonable to ask whether locomotion has matter, in anything like the sense that a stone, or a spark, *has* matter. Natural motion would not take place were it not for the presence of matter—earthy or fiery, as the case might be. But must this downward (or upward) *motion*, as such and considered in its own right, *have* matter? Motion as a case of "things that have no matter" would be exempt from Aristotle's restriction in his *Metaphysics*, and he appears here to have so regarded it, precisely where the concept of a motive power—the nearest thing to "force" included in his physics—was subjected by Aristotle to mathematical rules.

Returning now to the ancient dynamic theory of fall, we see that

---

11 δύναμις. Although in the Oxford Aristotle the English "force" is given here, Greek βία very seldom is found in Aristotle. As remarked above, force had the connotation of being a supernatural intervention, not a proper physical agency in natural phenomena.

12 Aristotle, *Physics* 7:5, 249b 25–29

Hipparchus did not discuss or compare speeds in fall through different media, at any rate in the summary by Simplicius. Hence he made no claim that speeds in fall by different weights of the same material are literally equal—and if he had done so, that would certainly have called forth comment by Simplicius. That claim is not found before the 16th century, though when it was made it had an ostensible basis in antiquity, as will be seen.

Either directly, or from the summary account by Simplicius, or by the critique of Philoponus, there is fair evidence that the dynamic theory of fall became widely known, passing first to the Arabs and from them into medieval Latin natural philosophy. The Arabic word *mail* for disembodied forces of this kind is reported by historians of medieval science, and the leftover force became *vis derelicta* in the writings of Francesco di Marchia during the 1320's. His account of fall soon gave way to the impetus theory of acceleration in fall as formulated by Jean Buridan at Paris in the mid–14th century and given mathematical expression by Albert of Saxony. That truly marked the introduction of dynamic thought into Aristotelian physics—not just its survival in a doctrine contradicting Aristotle's. The most orthodox Aristotelian natural philosophers remained hostile to impetus theory.

It is this uneasy symbiotic relation between kinematic and dynamic theories of motion, one arising to amend a previous form of the other, that continued to recur at each principal epoch in the history of physics, until conflicts disappeared with Newton's mathematical definition of force impressed upon a moving body. Even that victory of dynamic theories of motion may not have been final, in view of Einstein's principle of equivalence.

3

From 1300 to 1500

Early in the 14th century a significant new development in mathematics began with the work of Thomas Bradwardine of Merton College (Oxford) on speeds in motion. This was virtually coeval with the advent of dynamic thought in physics signalled by the *vis derelicta* of Francesco di Marchia in Italy, Bradwardine having composed it in 1328. Yet there was no direct connection between the two events, for Marchia concerned himself with the phenomenon of fall, whereas Bradwardine was motivated by a passage in the *Physics* of Aristotle that had nothing to do with falling bodies. For the passage cited in Chapter 2, giving Aristotle's rule for proportionalities in forced motions, continued as follows:[13]

> Again, if a given power move a given weight a certain distance in a certain time, and half that distance in half the time, half the motive power will move half the weight that same distance in the same time. Let E represent half the motive power A, and Z [be] half the weight B; then the ratio between the motive power and the weight in the one case is similar and proportionate to[14] that ratio in the other, so that each power will cause the same distance to be traversed in the same time.
>
> But if E move Z a distance Γ in a time Δ, it does not necessarily follow that E can move 2Z a distance ½ G in the same time [Δ]. If A move B a distance Γ in a time Δ, it does not follow that E, being ½ A, will in the time Δ, or in any fraction thereof, cause B to traverse a part of Γ whose ratio to the whole of Γ is proportionate to [i.e., the same as] the ratio between A and E, whether E be ½ or any other part of A; in fact, it might

---

13 Aristotle, *Physics* 7:5, 249b 30–250b 15

14 The redundant phrase suggests that in the old theory of proportion known to Aristotle, the definition of proportionality given by Euclid as "sameness of ratio" was not yet in use. The old, purely arithmetical (or number-theoretic) theory is set forth by Euclid in Book VII, while this definition is given in Book V.

well cause no motion at all...

For otherwise, Aristotle went on, one man might move a ship that in fact it would require many men to move, inasmuch as their combined power, and the distance they move the ship, are both divisible into as many parts as there are men.

In modern academic history of science it has become usual to say that mathematical physics is equivalent to an abandonment of Aristotelian natural philosophy in favor of Platonism, a great intellectual revolution which had to await the 17th century and whose leading exponent was Galileo. However, no historian has found it necessary to brand Bradwardine, or any of the medieval scholars to be mentioned in this section, a Platonist. Centuries before Galileo it had been possible to create a whole new theory of proportionality, to lay the mathematical basis for a whole new theory of fall, and devise a mathematical approach to the science of physics, all without forsaking Aristotle. To the contrary, those achievements began from Bradwardine's attempt to understand Aristotle's thought behind the text cited above.

The puzzle Bradwardine set out to solve was how Aristotle's rule of proportionality in motions could be reconciled with the case in which no motion results from application of a power to a weight. He was far from being the first to attack this problem, so he began by showing why no solution previously offered by philosophers could represent the thought of Aristotle. He then set forth his own solution:[15]

> The ratio of speeds in motions follows the ratio of the motive powers to resistances, and conversely. Or, to put the same thing in another way, the ratios of motive powers to resistances, and the speeds in motions, exist in the same order of proportion. And this is to be understood as geometric proportionality.

---

[15] H. Lamar Crosby, Jr., *Thomas of Bradwardine his* ***Tractatus de Proportionibus...***(Madison 1955), p. 112 gives the original Latin, with a very free translation on p. 113.

This rule, which Bradwardine believed to reflect accurately the thought of Aristotle about proportions in forced motions, has come to be called "Bradwardine's function" by those who express it symbolically[16] as $\left(\frac{F_2}{R_2}\right) = \left(\frac{F_1}{R_1}\right)^{\frac{v_2}{v_1}}$ —a translation into our mathematical symbolism that is as unlikely to represent the thought of Bradwardine in 1328 as his rather turgidly expressed conclusion is unlikely to have represented Aristotle's thought seventeen centuries earlier.

Bradwardine's thought was that *speeds* were so related to *net forces*[17] that speed (and hence motion) must vanish entirely when the power acting is not greater than the resistance to motion. If this terminology is strange, so was Bradwardine's conclusion strangely expressed—not because his thought was unclear, but because the language in which he was obliged to state it was as yet hardly equal to the task. Mathematical symbolism of any kind was not available to him; not even the writing of expressions in equation form had been invented. Neither equations nor symbols can easily represent the restrictions imposed by the rules[18] of proportionality before those rules were ignored in algebra. The thought of Bradwardine can be reflected properly only in words.

As to the function-concept, that entered into mathematics after symbolism took over in the 17th century. For Bradwardine's thought, the positive integers alone sufficed, and since the only cases dealt with

---

16 Cf. Clagett, p. 438.

17 I adopt this term to replace Bradwardine's "ratios of motive powers to [the related] resistances [against motion.]" In the symbolism of "Bradwardine's function" that was represented as a ratio of force to resistance encountered in a moving weight, F/R.

18 Proportionality was defined as "sameness of ratio," and ratio was defined as a relation in respect of greater or less between two magnitudes "of the same kind." Hence ratios of the sort created in algebra when a term is transferred from one side of an equation to the other were simply not allowed.

were those in which $R = 1; \frac{v2}{v1} = 2, < 2$, or > 2, it is misleading to use the term "function." Yet his thought had some interesting mathematical implications, and his approach led others in the 14th century to devise a wholly original theory of proportion which culminated in the earliest treatment of what are now called fractional exponents, and were christened "ratios of ratios" by Nicole Oresme, who completed that treatment.[19]

Although the medieval theory of proportion was an original contribution to mathematics, it added nothing to physical science in the sense of knowledge of natural phenomena. Bradwardine's rule did not correctly relate any forces, resistances, and speeds. Yet his manner of approach to mathematical physics proved to be remarkably fertile at the hands of others at Merton College, and later at Paris. This method was to devise and fit mathematical terms and reasoning with physical propositions. Not proofs, but numerical exemplifications (or later, geometrical diagrams) were used to verify their consistency. There was as yet no thought of utilizing measurements of physical phenomena, although the term "measure" was used in its mathematical sense, and was a concept that had underlain all earlier theories of ratio and proportion.

The most significant medieval application of the concept of measure (without actual measurement) was in the definition of the term "speed" as applied to an accelerated motion from rest. The concept of "speed" had remained undefined (unlike that of "equal speed.") Aristotle had offered only a definition of "swift" as applying to motion through a great distance in a small time; "swiftness," or speed, could not yet be mathematically defined.[20]

---

[19] For a discussion of some mathematical implications of the medieval theory of proportion and their symbolic expression, see my "Bradwardine's function, mediate denomination, and multiple continua," *Physis* 12:1 (1970), 51–68.

[20] We now think of speed as a "ratio" of distance to time, but in antiquity no ratio could exist except between two things of the same kind.

William Heytesbury of Merton College was probably the first to propose that speed during any completed "uniformly difform motion"[21] from rest be assumed equal to the speed at the middle instant. That device, appropriately named the "Merton rule," is loosely called the "medieval mean speed theorem." It was not a theorem, but a postulate—that speed of a completed motion from rest under uniform acceleration is measured by the speed in some uniform motion which carries a body an equal distance in the same time. Since what was taken as the measure was a certain constant speed, postulated to exist at the middle instant among differing degrees of speed, Heytesbury's rule is most fittingly described as the medieval middle-degree postulate. Its application was by no means restricted to locomotion, but also extended to κίνησις in general and was frequently used in discussions of heat, charity, and other qualities subject to increase or decrease. Nonetheless it was a significant contribution to define mathematically any concept having a physical application.

The middle-degree postulate was, in fact, never applied to motion in fall from rest during the 14th, or even in the 15th century, though many imaginary motions[22] were invented to fit the definition of uniformly difform motion. The only known statement that fall is a case of uniformly difform motion came in the mid- 16th century, without any accompanying evidence in its support.[23] Inasmuch as fall is uniformly accelerated motion, this historical situation was long puzzling to historians of science. But there was no way in which it could have been known that fall from rest is uniformly accelerated, without some ac-

---

[21] Uniformly difform motion was the medieval term for uniformly accelerated or decelerated motion, conceived either as continuous in the mathematical sense or as discretely incremental. See Clagett, Ch. 4.

[22] As for example, the motion of a point on the rim of a rotating wheel of ice, in an oven which caused the radius of the wheel to increase by thermal expansion at exactly the rate at which ice melted from the rim.

[23] The writer was a Spanish theologian, Domingo de Soto; see Ch. 4, and W.A. Wallace, "The enigma of Domingo de Soto," *Isis* 59:4 (1968), 384–401.

tual measurements of the physical phenomena—which, being merely a kind of observation, could shed no light on the hidden cause of motion and accordingly had no place in Aristotelian natural philosophy.[24]

Now, the causal explanation of fall offered by Marchia did not imply uniform acceleration during fall; rather, the falling body reached uniform motion after an initial acceleration. Even the new causal account to be discussed presently, developed a bit later in the 14th century, appeared to contradict one implication of the Merton rule, as that had been set forth by Heytesbury:[25]

> With respect, however, to the distance traversed in a uniformly accelerated motion commencing from zero degree and terminating at some finite degree, it has already been said that the motion as a whole, or its whole acquisition, will correspond to its middle degree... From the foregoing it can be sufficiently determined, for this kind of acceleration or deceleration, how great a distance will be traversed, other things being equal, in the first half of the time, and how much in the second half. For when the acceleration of a motion takes place uniformly from zero degree to some degree, the distance it will traverse in the first half of the time will be exactly one-third of that which it will traverse in the second half of the time.

It was Jean Buridan who set forth the theory of *impetus*, as an impressed force which did not diminish of its own accord (like that of Hipparchus, the Arab philosophers, and Marchia), but only when there was external resistance to motion or internal conflict with some contrary tendency to motion. About the middle of the 14th century Buridan wrote:[26]

> The first [conclusion] is that this impetus is not the identical local motion in which the projectile is moved, because this impetus moves the projectile, and the mover produces motion; therefore, the impetus produces that motion, and the same thing cannot produce itself.
>
> The second conclusion is that this impetus is not a purely successive thing, since motion is just such a thing and the definition of motion [as a successive thing] is fitting to it, as was stated elsewhere. And now it has

---

[24] Still less would actual measurement have any place in Platonist doctrine, for Plato discouraged attention to the inconstant and defective world of immediate sensible experience.

[25] Clagett, p. 272

[26] Clagett, pp. 536–7

just been affirmed that this impetus is not the local motion. Also, since a purely successive thing is continually corrupted and produced, it continually requires a producer. But there cannot be assigned a producer of this impetus which would continue to be simultaneous with it.

The third conclusion is that this impetus is a physically permanent thing distinct from the local motion in which the projectile is moved. This is evident from the two above conclusions and from what went before. And it is probable that this impetus is a quality physically present and predisposed for moving a body in which it is impressed, just as it is said that a quality impressed in iron by a magnet moves the iron to the magnet. And it is also probable that just as that quality—impetus—is impressed in the moving body along with the motion, by the movent, so with the motion it is remitted, corrupted, or impeded by resistance or by a contrary inclination.

Having closely associated impetus with speed, Buridan saw a case in which the speed of a natural motion could inaugurate the process that yielded a causal account of acceleration in fall:[27]

One must imagine that a heavy body not only acquires motion unto itself from its principal mover, i.e. its heaviness, but that it also acquires unto itself a certain impetus with that motion. This impetus has the power of moving the heavy body in conjunction with the permanent natural heaviness. And because this impetus is acquired in common with the motion, the swifter the motion is, the greater and stronger is the impetus. Therefore the body is moved from the beginning [of fall] by its natural heaviness only, and hence is moved slowly. Afterward it is [still] moved by the same heaviness and [also] by the impetus acquired at the same time [as before]; hence it is moved more swiftly. And because the movement becomes swifter, the impetus also becomes greater and stronger, and thus the body is moved by its natural heaviness and by that greater impetus, and so again it is moved faster; and in this way it will always and continually be accelerated, to the end.

Impetus theory removed the previous temporary character of acceleration in fall inherent in the *vis derelicta* account. But fall did not thereby become associated with uniformly difform motion, because Albert of Saxony soon proceeded to formulate the impetus theory of fall mathematically, as Buridan had not done. Albert examined the various ways in which growth of speeds and distances with time from

---

27 Clagett, pp. 551–2

rest might occur and eliminated those in which, contrary to Aristotle, infinite speed might be reached in finite time or over a finite distance. One way that Albert eliminated from consideration was this:[28]

> Natural motion does not accelerate by double, triple, and so on, in such a way that in the first proportional part of an hour it is a certain speed, and in the second proportional part of the hour [it is] twice as fast, and so on...

Because only completed motions were ever considered, it was customary to take proportional parts by halving repeatedly from the end of motion, making the first proportional part above a half-hour, the second, the preceding quarter-hour, and so on. In that manner, no first instant of motion would ever be reached, in accordance with a principle of Aristotle's that anything moving must have been already in motion, at any point selected. But an actual infinitude of speeds would surely be implied, each greater than the preceding, if motion could be divided back to its very beginning. Hence if the speeds increased as 1, 2, 3, 4, ..., the final speed in a finite motion could be infinite. Discarding all such progressions, Albert concluded as follows:[29]

> When some space has been traversed, [the speed] is some amount; and when double space is traversed, it is faster by double; and when triple space is traversed, it is faster by triple, and so on beyond...

Although Albert here intended successive motions, in order,[30] historians have read this as if each motion named were supposed to begin from rest, and concluded that Albert believed speeds in fall to be proportional to distances traversed from rest. But if that had been his conclusion, it would have been very easy for him simply to have said so—much simpler than to write out the above statement, and much

---

[28] Clagett, pp. 565–6

[29] Clagett, p. 566

[30] His purpose was to describe the first three steps in a single fall, from which anyone could go on, not to describe the first step in each of three different falls. The same format was used in attacks against the times-squared law three centuries later.

clearer to any reader.

In Buridan's account of fall there is a first motion during which only heaviness acts, necessarily a uniform motion at some speed.[31] That is followed by a second motion in which heaviness acts in conjunction with impetus acquired *simul* (all at once) at *its* beginning, and Albert regarded that as acting throughout this second motion, being uniform in speed and producing *simul* a still greater speed to act in the third motion, but not before. That succession of *uniform* speeds, each greater than the preceding, is in fact a *uniform* acceleration, but not mathematically continuous acceleration; it amounts to an acceleration by quantum-jumps of speed, so to speak. Albert might have written, more clearly for us now (but as something self-evident to his medieval readers) the following description of fall:

> When some space has been traversed, [the speed] is some amount; and *then* double space is traversed, *and* is faster by double; and *then* triple space, *this being* faster by triple, and so on beyond...

Under that rule, the successive time-intervals automatically become equal (not divided into proportional parts), while speeds are not increased proportionally either to the times (as they are in actual fall) or to distances from rest (as Albert is wrongly said to imply.) In his mathematization of the impetus theory of fall, times increase as the natural numbers, while the distances from rest accumulate at the ends of successive equal time as do the "triangular numbers" 1, 3, 6, 10, ... and not, as in actual fall, as do the square numbers 1, 4, 9, 16, ...

Albert's rule was easily grasped and it appears to have gone unquestioned before the 16th century. Its implication that in the second of two motions from rest, double the distance of the first motion is traversed, contradicted Heytesbury's rule for uniformly accelerated motion. In other words, the new causal explanation of acceleration during fall excluded fall from the category of uniformly difform motions. That in turn explains why, for 200 years, medieval natural philosophers did not debate whether speeds in fall were proportional to elapsed times, or to distances traversed from rest. Such a question, which may seem natural to historians of science now that laws of nature permeate sci-

---

31 Since only a single cause acts, its effect must be uniform. In the same way, when impetus is added, it adds uniform speeds in proportion to the degree of impetus added each time.

entific thought, would simply not have occurred to scholars at a time when causal explanation of natural phenomena was the sole purpose of science.

4

The Sixteenth Century

As commerce with the Near East revived, classical Greek texts in mathematics and science became known in Europe, mainly through Latin translations of Arabic versions, not from Greek. As universities began to be founded, Euclid's *Elements* in one Latin version became the standard text, with commentaries by Campanus of Novara. Whether the Arabic text he used had been responsible for the corruption of Book V is not known, though it seems probable, because the same faults are found in an earlier Latin translation by Adelard of Bath from an Arabic version in Spain.

Except for Book V, both Latin versions agreed fairly closely with the authentic *Elements*, in which Book V contains the general theory of ratios and proportionality for continuous magnitudes. In the standard medieval version, Book V was transformed into a largely superfluous treatment of continued proportion, applicable to numbers and discrete quantities, but inadequate for dealing generally with mathematically continuous magnitudes.

This transformation appears to have been deliberate, not a product of scribal carelessness or incompetent translation. One definition, essential to the general theory, was omitted, while another definition was inserted, not in place of the one omitted, but following another crucial definition and rendering it merely redundant. Apart from this tampering with two definitions, the rest of Book V remained unaltered (but essentially useless.) The omitted definition (Def. 4) was given by Eudoxus of Cnidos while Aristotle was still living;[32] Def. 5, in my opinion, was added by Euclid a century later. The medieval Book V removed those ancient advances, perhaps the greatest of all, returning

---

[32] It is extremely unlikely, however, that Aristotle was aware of the Eudoxian treatment of proportionality, even though he knew and adopted the system of homocentric spheres devised by Eudoxus.

the theory of ratio and proportion to its primitive state, a doctrine of numbers and rational fractions dating back to the Pythagoreans. As a result, medieval European mathematicians could not simply take up the subject at the point where Euclid and Archimedes had left off in Greek antiquity. That did not become possible until the mid–16th century, not long before Galileo was born.

The scientific Renaissance began in the 15th century after Cardinal Bessarion's deposit of classical Greek manuscripts in the library of St. Mark's at Venice. Printing from movable type began shortly afterward. The medieval Euclid had been printed twice when, in 1505, Bartolomeo Zamberti published at Venice a Latin Euclid based on the Greek text which included the complete and correct definitions belonging to Book V. That made rigorous treatment of proportionalities among mathematically continuous magnitudes (such as distances, times, and speeds) possible. But the hold of medieval tradition was so strong in universities that at least two generations elapsed before academic instruction in mathematics recognized the significance of the authentic text. Zamberti himself was partly responsible for the delay. He appears to have been less a mathematician than a humanist whose zeal was to remove the taint of infidel Arabic intervention from the classics of Greek learning. Instead of writing commentaries explaining the restored definitions of Book V, Zamberti pointed to them as examples of the errors of medieval translators. It remained for mathematicians to discern and explain the treasure that had lain unnoticed for a millennium.

The self-taught mathematician Niccolò Tartaglia, of Brescia, ultimately did that in 1543. He translated both Latin versions of the *Elements*, with the commentaries of Campanus, into Italian. His commentaries on Book V—the first to have been written by a first-rate mathematician since antiquity—restored the classic theory of proportion for continuous magnitudes in general.

As an experienced teacher of mathematics to private pupils having common sense and practical interests, Tartaglia knew how to bring Euclid—and the whole of mathematics—within easy reach for any reader of average intelligence, as was promised on the very title-page of this first living-language text of Euclid ever to be printed. That, however, did not change anything in the universities of his time, where medieval proportion theory continued to prevail for another half-century throughout Europe. The long-omitted definition was simple enough, but it had profound mathematical implications. Following the definition of "ratio" as a relation in respect of greater or lesser between two magnitudes of the same kind, it reads:

> Magnitudes are said to *have a ratio* to one another which are capable, when multiplied, of exceeding one another.

Tartaglia commented first that from this it follows that the circumference of a circle has a ratio to its diameter, whether or not that ratio can be represented by numbers (since any diameter multiplied by 4 certainly exceeds the circumference). Moreover, the same definition excluded from the domain of "ratio" both the infinite and the null, which remain unaltered by multiplication. Tartaglia did not say so, but it had been a basic principle of Aristotle's that there can be no ratio between the curved and the straight. A belief so old as to be found in Ecclesiastes I, 15 ("the crooked cannot be made straight") cannot be contradicted without involving profound consequences. For mathematics, one consequence was the rectification of curves; for physics, the abandonment of such Aristotelian dogmas as that of a necessary gulf between circular celestial and straight elemental motions.

Tartaglia's destruction of the spurious definition in the Campanus Euclid, and his restoration of meaning to the celebrated definition of "same ratio" that had been obscured, began from his perception that Campanus himself must have had before his eyes a valid ancient commentary (by Theon of Alexandria), but had been unable to understand it, or he would not have mingled it with meaningless distinctions in his own commentaries. This shows the acuity of Tartaglia (who did not read Greek) as a mathematician. He saw that part of the commentary was ancient, because it made mathematical sense of Euclid's definition, something that no one had managed to do in the Middle Ages. In his edition of Euclid, Tartaglia moved the commentaries explaining medieval proportion theory from Book V to the later arithmetical books, where they belonged (if anywhere.) Even in the 1570 English translation those were still published with Book V, as they continued to be in 1578 Latin textbook version by Christopher Clavius, standard in most universities throughout Europe during Galileo's lifetime.

Fortunately for modern physics, Galileo began his study of mathematics not in a university but under Ostilio Ricci, said to have been a pupil of Tartaglia's. He used the Italian version, rather than any Latin edition lacking Tartaglia's commentaries. Those are still a delight to read, even today, for their clarity, mathematical intelligence, and irreverent exposure of nonsense masquerading as scholarship.

Tartaglia's first published book had appeared in 1537 at Venice, where he taught mathematics privately to the end of his life. Titled *Nova Scientia,* that book dealt mathematically with artillery problems, such as the angle of elevation at which the maximum horizontal projection is obtained. Tartaglia's *Nova Scientia* finding of 45° was correct, but without the law of fall he was unable to correct the medieval conception of projectile paths—that the shot travels straight until the force projecting it is overcome by the natural tendency of things downward, weakened by conflict with that.[33] The projectile was then supposed to fall vertically to earth, as no body could obey two tendencies to motion at the same time, in Aristotelian physics.

Tartaglia remarked that he assumed straight motions only as a practical approximation, for the path must be curved at every part once the ball has left the gun. He seems not to have been familiar with impetus theory, as he never used the word *impetus*. Neither did Tartaglia give any account of acceleration in fall, beyond his likening that to the hastening of a traveller's steps as he neared home.

Impetus theory seems to have fallen into neglect also among natural philosophers in the early 16th century, unlike medieval kinematics. By 1550 the first known statement had been ventured that fall is an example of uniformly difform motion. The author was Domingo de Soto, who offered no evidence for his statement and appears to have been unaware that it conflicted with Albert of Saxony's mathematization of impetus theory. Father William A. Wallace, who undertook to explain the "enigma of de Soto" and was thus enabled to recognize the classificatory purpose behind the emergence of uniformly difform motion, concluded that the unique statement by de Soto was made mainly because his superiors in the Dominican order disliked mathematical abstractions. They wanted physical examples for things taught in natural philosophy.[34]

---

[33] On Aristotelian principles of physics, only the stronger of two tendencies to motion could be obeyed by a heavy body, and the inherent tendency downward could not be weakened, being natural.

[34] W.A. Wallace, "The concept of motion in the sixteenth century," *Proceedings of the American Catholic Philosophical Association*, Washington, 1967.

In 1551 Tartaglia published a book on the raising of partly sunken ships, often menaces to navigation in the Venetian Gulf. His method used the Archimedean principle, and on the first page he remarked that the speeds of bodies sinking in water were as their specific gravities. Archimedes had not discussed speeds of sinking or rising in water, but Tartaglia repeated his remark in a section of the book presenting, in Italian translation with dialogue commentaries, Book I of Archimedes' *On bodies in water*.[35] The great importance of Tartaglia's remark for new 16th-century developments in the history of fall is not widely recognized.

One of Tartaglia's pupils in 1547 had been G.B. Benedetti, who first asserted, and offered a proof for, the proposition that bodies of the same material but different weights fall with equal speed through the same medium. That first appeared in 1553, in the dedicatory letter to Benedetti's first book, with the promise of further physical propositions to follow. The book concerned solution of all problems in Euclid with a ruler and a compass of fixed setting.[36] Benedetti remarked that the place was not appropriate for his proposition on fall, but said that he did not want to risk his priority. Hence his discovery had been very recent, not earlier than 1552, when Benedetti had indeed good reason to fear that others might hit upon it soon.[37]

There is little doubt that Benedetti, reading Tartaglia's 1551 book on the raising of sunken ships, noted the remark that bodies sink in water

---

35 N. Tartaglia, *Regola generale...*(Venice 1551), ff. 1*v* and English translation in T. Salusbury, *Mathematical Collections and Translations* (London 1665, repr. 1967), vol. 2, pp. 484, 335, Salusbury having printed the Archimedean work with Tartaglia's commentaries separately, and first.

36 This was a celebrated problem in mathematics at the time, to which Tartaglia had devoted much study. Benedetti's analysis showed him to be an original and talented mathematician.

37 Tartaglia's solution of the cubic had been published by Girolamo Cardano despite his vow to keep it secret until Tartaglia made it public. Even if Benedetti had not known this, Tartaglia printed the relevant letters in an appendix to his 1551 book on raising sunken ships, the idea for which Cardano had also plagiarized in his *De subtilitate* of 1550.

with speeds proportional to their specific gravities. No reason appeared why fall through air should not be subject to a like condition, defying Aristotle's rule that speeds were proportional to simple weights. Benedetti's proof applied the Archimedean principle, also to be found in Tartaglia's 1551 book. That made it not unlikely that before long, others would reach the same conclusion or would hear of Benedetti's discovery and claim it for themselves. Hence his haste to publish it.

Announcement of equal speed in fall for bodies of different weights created controversy, in which some seem to have disputed this because it contradicted Aristotle, while others cited other authorities. At any rate in 1554 Benedetti published a book on speeds in fall written "against Aristotle and all the [natural] philosophers."[38] Later he was invited to lecture at Rome on his novel ideas. There he was heard by a Belgian, Jean Taisnier, who published as his own, at Antwerp in 1561, Benedetti's book of 1554. It was translated into English as Taisnier's, and by the 1570's the idea of equal speeds in fall had spread to most of Europe, together with a mathematical demonstration in its support.

What had been lacking was any assertion that the proposition had been tested by observations. This was supplied by the Dutch engineer Simon Stevin in 1586, who had read the Belgian edition. With a friend, Jan de Groot, Stevin compared times of fall from a height of thirty feet and found them equal,[39] but he also saw an error in the proof, for which he believed Taisnier responsible. The mistake had been Benedetti's, who had already detected and corrected it in a second edition of the book plagiarized by the Belgian, published later in 1554.

The fact that no one seems to have put so novel a statement to actual test before Stevin, long after it had circulated very widely, is a commentary on the natural philosophy that preceded scientific physics in the modern sense. It was not a difficult experiment to carry out, but

---

[38] G.B. Benedetti, *Demonstratio proportionum motuum localium...*(Venice (1554); English translation in S. Drake and I.E. Drabkin, *Mechanics in Sixteenth-Century Italy* (Madison 1969), pp. 154–65.

[39] Simon Stevin, *De Beghinselen der Weeghcoonst* (Leyden 1586); see E.J. Dijksterhuis, *The Principal Works of Simon Stevin*, vol. 1 (Amsterdam 1955).

to mathematicians as well as natural philosophers a test appeared unnecessary; only reasoning, or the authorities adduced, really mattered. Stevin was a first-rate mathematician, the inventor of decimal fractions and the first to break with ancient arithmetical theory (and with Euclid himself) by declaring in print that one is a number.[40] But Stevin was not only a mathematician; he was also, and primarily, an engineer no less interested in practice than in theory.

In 1586 Galileo was tutoring privately in mathematics, as Tartaglia had done most of his life. For the text of Archimedes he must have been using the 1543 edition by Tartaglia, for that alone included the work on bodies in water on which Galileo based his own first scientific essay, *La bilancetta* of 1586. Now, in that year Galileo also began his first manuscript on motion, a pupil-teacher dialogue of the kind Tartaglia had used for his commentary on Archimedes in 1551, and previously in 1546 when he edited and commented on the principal medieval work on statics, attributed to Jordanus Nemorarius.[41]

Galileo's dialogue on motion included the proposition that bodies of the same material fall through the same medium with equal speeds, regardless of their weights.[42] The mathematical demonstration he offered was similar to Benedetti's a generation earlier, and some other arguments in the dialogue also resemble those of Benedetti in his second book, newly reprinted in 1585.[43] Many think that Galileo's source for equal speed in fall had been Benedetti, but more probably it was the very same source as that of Benedetti; namely, Tartaglia's book on raising sunken vessels, which had been reissued several times, bound with other works of his. If so, that book inspired two attacks against Aristotelian physics, a generation apart—without Tartaglia's ever having perceived the further implications of his own statement about speeds of sinking through water.

---

[40] Euclid defined number as "multitude of units," and since one is not a multitude, two was the smallest number in classical Greek mathematics.

[41] N. Tartaglia, *Quesiti, et inventioni diverse* (Venice 1546); see also his posthumous *Jordani opusculum de ponderositate* (Venice 1565).

[42] English translation by I.E. Drabkin in our *Galileo On Motion and On Mechanics* (Madison 1960) pp. 89–90.

[43] G.B. Benedetti, *Diversarum speculationum*...(Torino 1585).

One of many reasons for doubting that Galileo had seen the 1585 edition of Benedetti's works is that in that book Benedetti had set forth the same kind of explanation for acceleration in fall as that of medieval impetus theory. But Galileo, in his writings on motion during the 16th century, treated acceleration as *temporary* only, at the beginning of fall and as the result of loss of the impressed force that sustained it, or had moved it upward. That was precisely the explanation already given by Hipparchus in antiquity, and at least implied by Philoponus in the sixth century, which was once more proposed in the 1320's by Marchia. Galileo appears to have arrived at it independently, and to have been very proud of it. After he found it attributed to Hipparchus in a book by Benedict Pereira, Galileo noted that Hipparchus had neglected to explain more than the case of fall ensuing upon upward projection.[44]

Now, that was mistaken, but the mistake was due to failure on the part of Pereira to describe the work of Hipparchus as fully as Simplicius had done, writing:

> Hipparchus said that the cause of such events is that violent motion precedes natural motion; that in being moved naturally downward, first came motion and projection upward, so when returning downward, this force that projected upward is weakened and diminished, and near the end this force is broken and remitted so that swifter motion then exists.—But Hipparchus deserved reprehension by Alexander [of Aphrodisias], who said this may be true for natural motions when something violent immediately preceded, but cannot also hold in natural motions in which there was not some violence before...[45]

The fault thus charged by Alexander (as here cited) against Hipparchus was perhaps properly chargeable only against his 14th-century followers such as de Marchia; it did not go back to Greek antiquity. The case of fall from rest at a height is even more common, and physically more interesting, than the case of prompt return after projection upward. In his Pisan *De motu* of 1591–2 Galileo included a passage about this that deserves notice:[46]

---

[44] Drabkin & Drake, *op. cit.*. pp. 89–90.

[45] Benedict Pereira, *De communibus omnium rerum*...(Venice 1586), Bk. 14, p. 475.

[46] Drabkin & Drake, *op. cit.* n. xx, p. 109.

> Suppose that there are two bodies equal in size, one of wood and the other of lead; that the weight of the lead is 20, and of the wood, 4; and that both are held up by the line *ab*.

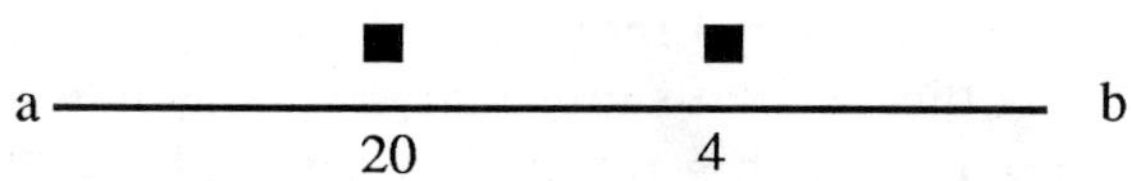

> Now in the first place, it is clear that these bodies press down with a force equal to that with which the line *ab* presses upward. For if they exerted more than that pressure, line *ab* would not hold them up, and they would move down in spite of the line. But actually they do not move downward in the air because they are not exerting weight upon the air, the medium through which they must move... It is rather on *ab* that the bodies are exerting pressure... But when they are released by the line they still retain, at the first point of their departure, an impressed contrary quality that impels them upward, and this quality is lost not instantaneously but gradually. The lead has 20 units of this contrary quality to be used up, and the wood, 4...
>
> In the time in which one unit of the quality departs from the lead, more than one unit has left the wood; and consequently while the lead has recovered only one unit of weight, the wood has recovered more than one unit. It is because of this that the wood moves more swiftly during that time... On the other hand, because the lead ultimately reacquires more weight than the wood, it follows that by that time, the lead is moving much more swiftly [than the wood.]

Galileo certainly did not get this erroneous but interesting analysis from Hipparchus; still less did it come from a medieval natural philosopher or mathematician. Written about the time of Galileo's famous demonstration of equal speeds in fall from the Leaning Tower of Pisa, it demolishes fanciful reconstructions of that event in which it is frequently said that one falling weight was leaden and the other wooden. The original account, written by Vincenzio Viviani in 1657, who had it from Galileo in 1640,[47] stated that both weights were of the same mate-

---

[47] At that time Galileo was blind and Viviani took his letters in dictation. A letter from the professor of mathematics at Pisa recounted recent experiments from the Leaning Tower, unaware of Galileo's much earlier demonstration there. That is why Viviani had the details right, though he was not born until long after the event.

rial, which in 1590 was the only case that had been considered by Benedetti, Stevin, or Galileo as having been mathematically established or verified by observation. From the above it is evident that Galileo might expect the lead weight to move much more swiftly than that of wood after the initial motion. The proof for equal speed of fall held only for bodies of the same material, and it would have been foolish for him to risk his reputation with his students by using weights that he would not expect to move with the same speeds.

That the lighter of two materials begins fall from rest more swiftly than the heavier was asserted more than once in *De motu*, and in very recent years such a possibility has been seriously investigated by physicists. Galileo said that he had observed the phenomenon, which is puzzling, though it was reported also in a book by one of his teachers at Pisa, Girolamo Borro, whose explanation of the alleged experimental fact Galileo did not accept.[48] Dr. Thomas Settle has reported that a person holding weights in both hands will release the lighter a bit sooner than the heavier, despite an intention of letting them go at the same time.[49]

Also of interest is Galileo's implication that the line *ab* pushed up with different forces against the two weights. The equality of action and reaction (Newton's third law of motion) was by no means a commonplace in 16th-century mechanics; indeed, the idea of a table-top pressing upward when a weight is placed on it would have been ridiculed by Aristotelian physicists, and even today seems odd to many beginning students. Galileo brought to his early studies of motion much that had not traditionally been considered relevant to such studies. Even at the outset, he thought more like later physicists than like the earlier natural philosophers whose authority counted heavily with philosophers of his own time. But his conviction that acceleration was a brief event at the beginning of fall, and that it could be ignored for convenience (as Tartaglia had neglected the curvature of paths of cannonballs), held Galileo back for many years. The conflict between ki-

---

[48] Girolamo Borro, *De motu gravium et levium* (Florence 1576), p. 232. Galileo had already mentioned this book on the first page of his 1586–7 dialogue on motion.

[49] T. Settle, "Galileo and early experimentation" in *Springs of Scientific Creativity*, ed. R. Aris, H.T. Davis, and R.H. Steuwer (Minneapolis 1983).

netic and dynamic theories of fall, seemingly resolved by medieval impetus theorists, came to life again with Galileo's Hipparchian preference.

The first unequivocal statement that speeds in fall are proportional to distances from rest appeared in 1584, one year before Galileo left the University of Pisa without a degree and began teaching mathematics to private students. The proposition appeared in *De motu tractatus* by Michael Varro, a Swiss merchant who had studied mathematics in his youth and aspired to imitate Archimedes in "demonstrating by reason and proving by experiment" as he said in the opening sentence of his dedicatory letter.[50]

Varro's terminology is entirely different from that of impetus theory. Natural force, by which a thing tends toward its natural place, moves it with continued and orderly motion. That is why it is easier to move a moving thing than a thing at rest. The same force pressing in motion moves a thing more greatly in motion than at rest, and the more greatly as the motion becomes greater. Speeds are greater in the proportion of distance moved to the whole space of the motion. Varro's line and triangle diagrams illustrating his proposition closely resemble those found among Galileo's working papers in 1602–4.

---

[50] M. Varro, *De motu tractatus* (Geneva 1584), f. A2. The statement on p. 12 *æqualia spatia ab eorum principiis numerata, æqualibus temporibus habebunt*, is exemplified by diagrams on p. 13.

5

Discovery of the Law of Fall

Galileo's working papers on motion from 1602 to 1637 are now in Volume 72 of the Galilean manuscripts at Florence. On *f.*107*v* of that volume is his first surviving record of precisely measured distances. They are in *punti*, of 0.94 mm. per *punto* as determined from notations on *ff*. 115, 116, and 166 of the same volume. A few days later these first measurements became linked with the law of fall, so we may begin from those. Calculations at the center of *f.*107*v* are of the form 60x + (60 -n), implying the use of a ruler finely divided into 60 equal parts.[51] Before outlining Galileo's experimental work in obtaining the numbers he tabulated on *f.*107*v*, the reason behind that work will be given.

Late in 1602 Galileo wrote a letter to his friend and patron Guidobaldo del Monte, author of the best 16th-century book[52] on mechanics, recommending the use of pendulums 4 to 6 feet long in experiments relating to descents along a circular arc. In that letter[53] he communicated two theorems concerning descents along chords to the

---

[51] The same ruler was used in drawing and measuring diagrams on several pages of the working papers during the years around 1604.

[52] Guidobaldo del Monte, *Mechnicorum liber* (Pesaro 1577); Italian translation by F. Pigafetta, Venice 1581; abridged English translation in S. Drake & I.E. Drabkin, *Mechanics in Sixteenth Century Italy* (Madison, 1969).

[53] Translated in S. Drake, *Galileo at Work* (Chicago 1978), pp. 69–71.

bottom of a circle, with his conjecture that times along arcs to the lowest point are equal for all lengths of arc.[54] In 1603, after deriving two least-time theorems for straight descents, Galileo saw that he could advance further only by some rule for increase of speed in straight natural descent from rest.

Aristotle had defined "greater speed" as the traversal of greater distance in the same time, so Galileo conceived the idea of equalizing several times during a relatively slow straight descent and taking the distances traversed as measures of speeds. A grooved inclined plane about 2 meters long (over 2,100 *punti*) was tilted 60 *punti*, to an angle of 1.7° with the horizontal. Along the groove a bronze ball descended repeatedly from rest while Galileo divided the time into 8 equal intervals, probably by singing at beats of a half-second each.[55] The place of the ball at each beat was marked; at those marks, strings were tied around the grooved plane. Probably gut strings were used, as when frets were tied around the neck of a lute—firmly, but capable of being adjusted when tuning the instrument. After the ball passed over each string, a faint bumping sound was audible as it struck the plane. Positions of the strings were patiently adjusted until all sounds coincided with notes of the tune, and the distance from resting contact of ball with plane to the lower side of each string was noted. Analysis shows that Galileo was accurate within $\frac{1}{64}$ second for every string except the lowest. Later he adjusted that string by about 20 *punti* (say 2 cm.), and also marked a + or a – sign after four of his original measures which seemed to Galileo a bit short or a bit long when he made subsequent

---

54 This conjecture was mistaken, as Galileo learned in 1604 when he began measuring times precisely. His knowledge of pendulum motions at each stage of the work will not be discussed in this monograph except as it relates to discovery of the law of fall.

55 Calculation shows the times to have been very nearly 0.55 seconds each. Marching songs usually have such a tempo, at which even a slight departure from the rhythm is easily detected.

re-runs (these then sounding to him early, or late.)

In equal successive time intervals, Galileo found, the speeds increased as do the odd numbers 1, 3, 5, 7, ... etc.[56] This simple and exact rule doubtless surprised Galileo as much as it delighted him, for only a year before he had said to Guidobaldo that when put to a test with material objects, rules deduced by mathematicians simply did not agree with observation. Material impediments, as Galileo called friction, air resistance, absence of perfect smoothness and hardness, or the like, played a very important part in his physics from 1602 through 1609.

Having found a simple arithmetical law for a natural motion, Galileo thought next of *measuring* short intervals of time instead of merely equalizing them. Near the bottom of *f*.107*v* he drew a diagram of the device that he used from then on for measuring the times of motions. In *Two New Sciences* he described its simplest form, in which flow of water from a large pail through a slender tube in the bottom was collected during each motion, was weighed on a sensitive balance, and the weights were treated as measures of the times.[57] The sketch on *f*.107*v* suggests (and the precision of Galileo's few surviving records of timings confirms) my belief that in fact Galileo built and used a somewhat more reliable and consistent means of starting and ending flows

---

[56] At this time Galileo could measure only overall speeds and did not yet have a way to *time* brief motions. Because the times had been equalized, the distances were direct measures of overall speeds. It did not occur to Galileo at first to consider them as simple *distances* also, so he did not sum the successive odd numbers and perceive the times-squared law at once.

[57] Galileo, *Two New Sciences*, tr. Drake (2d ed. Toronto 1989), pp. 169–70.

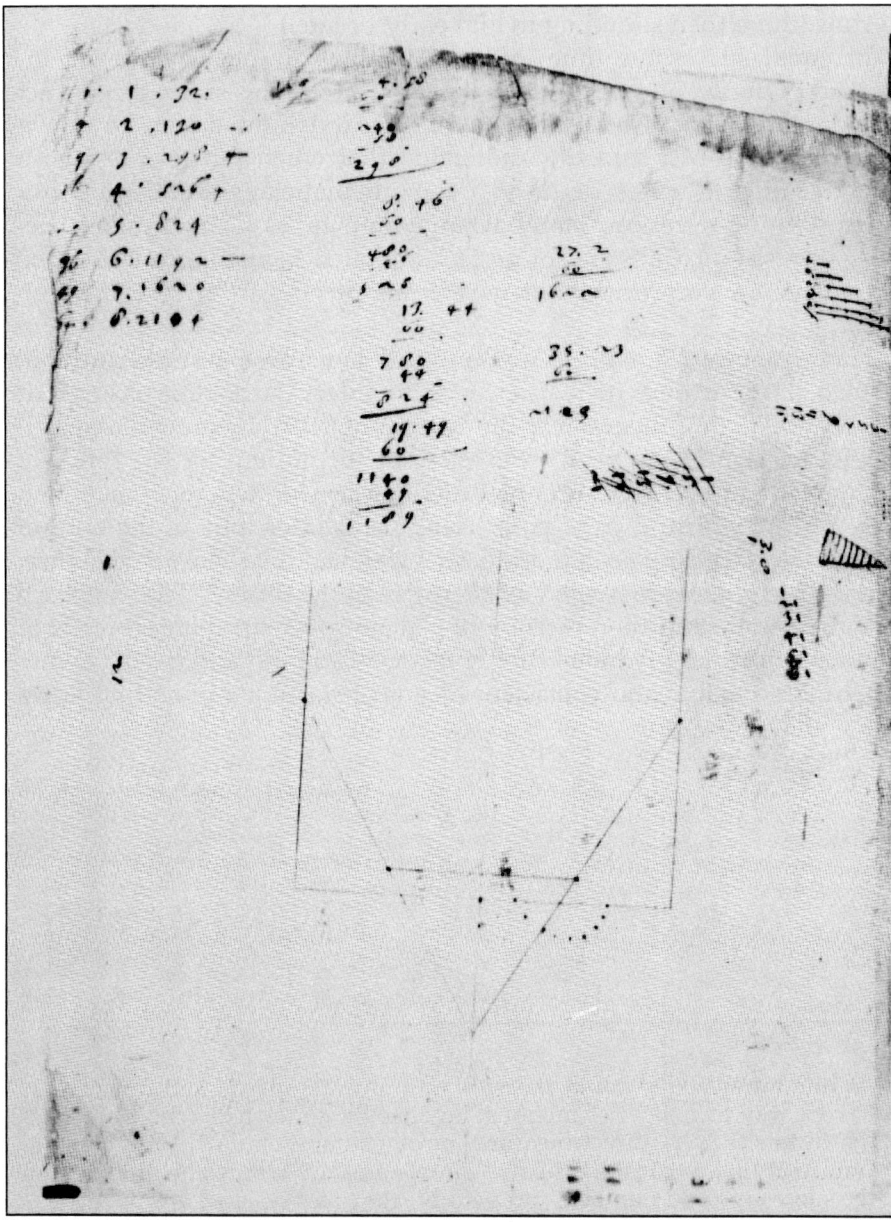

Galileo's earliest surviving measurements of accelerated motion, *f.*107*v*, volume 72, Manoscritti Galileiani in the Biblioteca Nazionale Centrale di Firenze (reproduced by permission). The squared numbers in the left margin are barely legible in reproduction (see text).

than mere removal and replacement of the thumb or a finger at the end of the tube. That simple procedure has, however, been found to give results twice as reliable as Galileo asserted in his later book, and his own extant timings confirm this.[58]

The rate of flow of water from Galileo's device was 3 fluid ounces per second, very nearly indeed. This round number in an old standard unit was mere coincidence, because Galileo was not attempting to measure time in any accepted unit, let alone in astronomical seconds. He always worked entirely in ratios and proportionalities, from which any unit of measurement would of course cancel out.

For practical reasons, Galileo did use a standard unit of weight; namely, the grain, of which there were 480 to the fluid ounce. Galileo's recorded times are in grains weight of flow of water from his timing device. Later he adopted as his unit of time the flow of 16 grains of water from that device; 16 grains weight is almost exactly 1 gram in c.g.s. units.[59] That unit will be called the *tempo* (plural, *tempi*). How Galileo came to adopt it will be explained below.

Galileo's first recorded timing was noted on *f.*154*v* in the form of a column of numbers, totalled as 1,000 + 107 + 107 + 107 + 16 = 1,337 (grains weight of water collected during a motion from rest.) That was his timing of fall through 4,000 *punti* = 376 cm. It was his least exact timing, being about $\frac{1}{30}$ second too high. From his collecting vessel Galileo first poured off 1,000 grains, probably into a container marked as holding that amount. He then marked on the collection vessel the level at which the remainder stood, and later on he used that mark as

---

58 Galileo said that in more than a hundred timings along inclined planes, agreement with the times-squared law had been found never to vary more than one-tenth of a pulsebeat. Dr. Thomas B. Settle reported in 1961 that with little practice he had achieved about double that precision. Cf. Settle, "An Experiment in the History of Science," *Science* 133 (1961), pp. 19–23.

59 Settle avoided the nuisance of repeated weighings by collecting flows in a graduated cylinder and taking 1 cc. as 1 gram. It is a curious fact that Galileo, though he did not have a graduated cylinder, used volumetric determinations in place of weighings by marking his collection vessel, as will be seen. That 1 gram of water represented 1 *tempo* for Galileo and was also the unit for Settle's ratios of times was purely coincidental, since the rates of flow were doubtless very different.

representing 320 grains (the rounded sum of 3 × 107.) The final figure, 16 (grains), was indirectly weighed; it represented the weight of the collecting vessel damp, less its dry weight. As a former medical student Galileo was familiar with that way of accounting for fluid that adhered to the sides of a vessel after pouring from it.

Directly beneath 1,337 Galileo entered 903, a second timing in grains flow during a motion from rest. That was his timing of fall from 2,000 *punti* = 188 cm. (about 6 feet), a very convenient height at which the timing with Galileo's device did not require difficult action on his part. (To dislodge a weight from double that height, as before, in exact coordination with the starting of flow, was less simple.) Timing of 903 grains for fall 2,000 *punti* is nearly exact; I calculate that 911 or 912 grains would be precise at the latitude of Padua (taking g = 980.7 cm/sec), and that Galileo's finding was little more than $\frac{1}{200}$ second too low. In this case there is also evidence that Galileo had marked the level at which 903 grains stood in the collection vessel before he weighed the water. That is suggested by the fact that there was no column of figures, totalled, as before (and as on *f*.189*v*1 a bit later), and it is confirmed by a number on *f*.151*v* which Galileo had obtained when he first set out to relate fall to the pendulum. In order to understand the initial steps, it should be mentioned that Galileo used swings of pendulums only through a small arc to the vertical, for a practical reason. The only swing that could be timed with precision was from the instant of releasing the bob to the sound of impact with a block that had been fixed against a side of the bob when hanging plumb.

On *f*.151*v* there are several diagrams intended to represent geometrically some distance of fall from rest and the length of pendulum swinging to the vertical in a related time. There is also a freehand sketch of two meshed gears, and the calculation 53 × 30 = 1,590. As already said, Galileo measured lengths with a carefully engraved ruler 60 *punti* long. He also habitually first eliminated the fraction ½ when performing a multiplication or a division. Thus the *f*.151*v* calculation stood for $26\frac{1}{2} \times 60 = 1{,}590$, the length in *punti* of the measured pendulum which, by my calculation, swings at Padua through a small arc to the vertical during the flow of 903 grains of water through Galileo's timing device—exactly the flow he had recorded for fall 2,000 *punti*.[60]

Now, the pendulum that takes the same time to the vertical as does fall through 2,000, in any units whatever and anywhere, is in fact (to the nearest digit) 1,621 units long, not 1,590. Galileo's small error in timing of fall through 2,000 *punti* had resulted in shortage of 31 *punti* for the pendulum that would time that fall exactly by swing to the vertical through a small arc. Although Galileo's two timings had been exactly consistent, the length-ratio $\frac{2{,}000}{1{,}590}$ (= 1.2579) was considerably above the correct $\frac{\pi^2}{8} = 1.2337$, or $\frac{2{,}000}{1{,}621}$ to four significant places.

Galileo's skill as an experimentalist is illustrated by the pendulum length which he recorded on *f.*151*v* as being 1,590 *punti*. That figure can have been obtained only by finding the pendulum whose swing to the vertical through a small arc accompanied flow of 903 grains weight of water through his timing device. Implied is his having started with a pendulum about 5 feet long, and then having patiently adjusted it until water flowed precisely to the previously marked level while the pendulum swung to the vertical. This procedure throws light on the sketch of two meshed gears, seen also on *f.*151*v* but appearing out of place on a page bearing diagrams relating fall to the pendulum.

When he had measured the pendulum for 903 grains flow, we might expect Galileo next to find the pendulum timed by 1,337 grains. That would be longer, and very inconvenient to alter in length repeatedly as before. Running the string over a nail in a movable upright and anchoring it to a bench would allow the nail to be raised and lowered by gears and a crank. Such a scheme accounts for the sketch. In the end, however, Galileo timed the pendulum for $\frac{1{,}337}{2} = 668\frac{1}{2}$ grains of flow

---

60 The precision of this measurement, though it was of no use for the problem Galileo was addressing, helps to explain the almost perfect values he had for the units described in the Epilogue. His *punto* of 0.94 mm should be 0.9422119204, and his *tempo* of 1/92 second should be 1/91.88024932 second, if my calculations are correct.

as being 870 *punti* in length, and then he timed the doubled pendulum of 1,740 *punti* at 942 grains of flow. At any rate that is what is implied by the working papers which Galileo preserved.

From the pendulum measurements Galileo now had at hand, a table of the kind shown below could be compiled. Though I doubt that Galileo troubled to form a table, mine will serve to show the way in

Length of Pendulum in *punti*	**Time to the Vertical in grains flow**
870	668 1/2
1,740	942
3,480	1,337
6,960	1,884
13,920	2,674
27,840	3,768

which he arrived at his discoveries by simply applying the theory of ratio and proportionality set forth in Euclid's *Elements*. My table has been extended far enough to show the source of a very important number found in two of Galileo's surviving working papers, on one of which he was writing when he first recognized the law of fall in times-squared form from its mathematically equivalent mean-proportional form.

The first column was formed by successive doubling, and the second by alternate doublings. Except for a slight discrepancy with the second timing, each column separately is accordingly in continued proportion. Galileo's new time unit, the *tempo,* came into being when he related the two columns horizontally, so to speak. His original measure of time in grains of flow through his timing device having been completely arbitrary, he was free to alter it in any ratio he pleased.[61] Taking each time to be the mean proportional between 2 and the length of pendulum, the two columns become related line by line. The same numbers result also, almost exactly, from division of each time in grains by 16, so 16 grains of flow became 1 *tempo*—the new unit adopted as a result of this application of the Euclidean theory of proportion.

Because division by 16 of the times in grains weight of flow did not *exactly* produce that mean-proportional relation between the two columns, Galileo made an adjustment which resulted in his changing 27,840, as seen above, to 27,834. That work was done on one of the working papers of which only a part survives. On the blank side he wrote, in 1609, a note on another topic, cut it out, and pasted it on *f.*90. That was lifted at my request, and on the hidden side I saw the number 27,834, twice, with enough words to identify it as a "diameter." Galileo's diagram and calculations were thrown away with the part of the page cut off, but they had been finished before Galileo discovered the law of fall, because 27,834 played a crucial part in the calculation on *f.*189*v*1 which put that discovery into Galileo's hands.

With his adoption of the *tempo* as the unit of time, Galileo had the pendulum law in a restricted form; that is, for any set of pendulums successively doubled in length. It would have been a difficult task to test it for successively tripled pendulums, let alone for any other integral multiples, and quite impossible to establish it experimentally in its complete generality. What Galileo did next is seen on *f.*154*v* (on which he had entered his first two timings.) He now calculated the mean proportional of 118 and 167—the times to the vertical, in *tempi*, for the two pendulum lengths 6,960 and 13,920 *punti*—getting 140 (*tempi.*) The mean proportional of those two lengths is 9,843, so if his restricted pendulum law were perfectly general, a pendulum 9,843 *punti* long would swing to the vertical in 140 *tempi*.

On the other side of the same page, *f.*154*r*, Galileo wrote the note *filo*

---

[61] The unit of length could have been altered instead, but Galileo had recorded previous careful measurements made with his ruler. It was easier to take a new time unit than to graduate another ruler.

br. 16—"the string is 16 *braccia* long." From two lines drawn and labeled by Galileo at Padua, I measured one *braccio* as containing about 620 *punti*. At 615 *punti* per *braccio*, length of the pendulum would be 9,840 *punti*, or about 30 feet. Such a pendulum could be hung from a window over the courtyard of the University of Padua and timed, protected from wind. At Padua it would reach the vertical through a small arc in 141 *tempi*, by my calculations. Thus Galileo was fully assured of the complete generality of his pendulum law in mean-proportional form that is mathematically the same as our law that periods of pendulums are as the square roots of their lengths.

In order to finish linking fall with the pendulum, Galileo needed one bit of information that has not yet been identified, the time of fall through a distance equal to the length of a timed pendulum. The length he chose was $2 \times 870 = 1{,}740$ *punti*, and that fall takes 850 grains flow of water. The figure 850 exists in Galileo's extant working papers, but only by implication. At top left on *f.189v1* there is a column of figures, added to total 1,988, in which two separate pendulum timings are recognizable; subtotals are added here in parentheses:

13 + 530 + 320 + 180 (=1,043) and 95 + 320 + 530 (=945), total 1,988.

The numbers 180 and 95 represented actual weighings of collected water, while the number 13 represented adjustment for unweighed water adhering to the sides of the collecting vessel. The other two numbers, 530 and 320, were weights in grains found by pouring water from the collecting vessel to marks previously made. The mark for 320 grains had been made during the very first timing, as noted earlier. Hence the mark for 530 grains had been placed when a timing of 850 grains was made, and 850 grains of flow does time the fall through 1,740 *punti*. Thus the column of figures totalled at the top of *f.189v1* recorded two separate timings of a pendulum of length 1,740 *punti*, one through a very wide and the other through a very small arc, doing this for a reason unrelated to his discovery of the law of fall.[62]

In the central calculation on *f.189v1*, the ratio used was time of pendulum 1,740 *punti* to time of fall through 1,740 *punti*, or $\frac{942}{850} = 1.108$

---

[62] Galileo was diverted into a problem he had considered in 1602, before he had any means of timing brief motions. Discussion of this would unnecessarily complicate the present monograph.

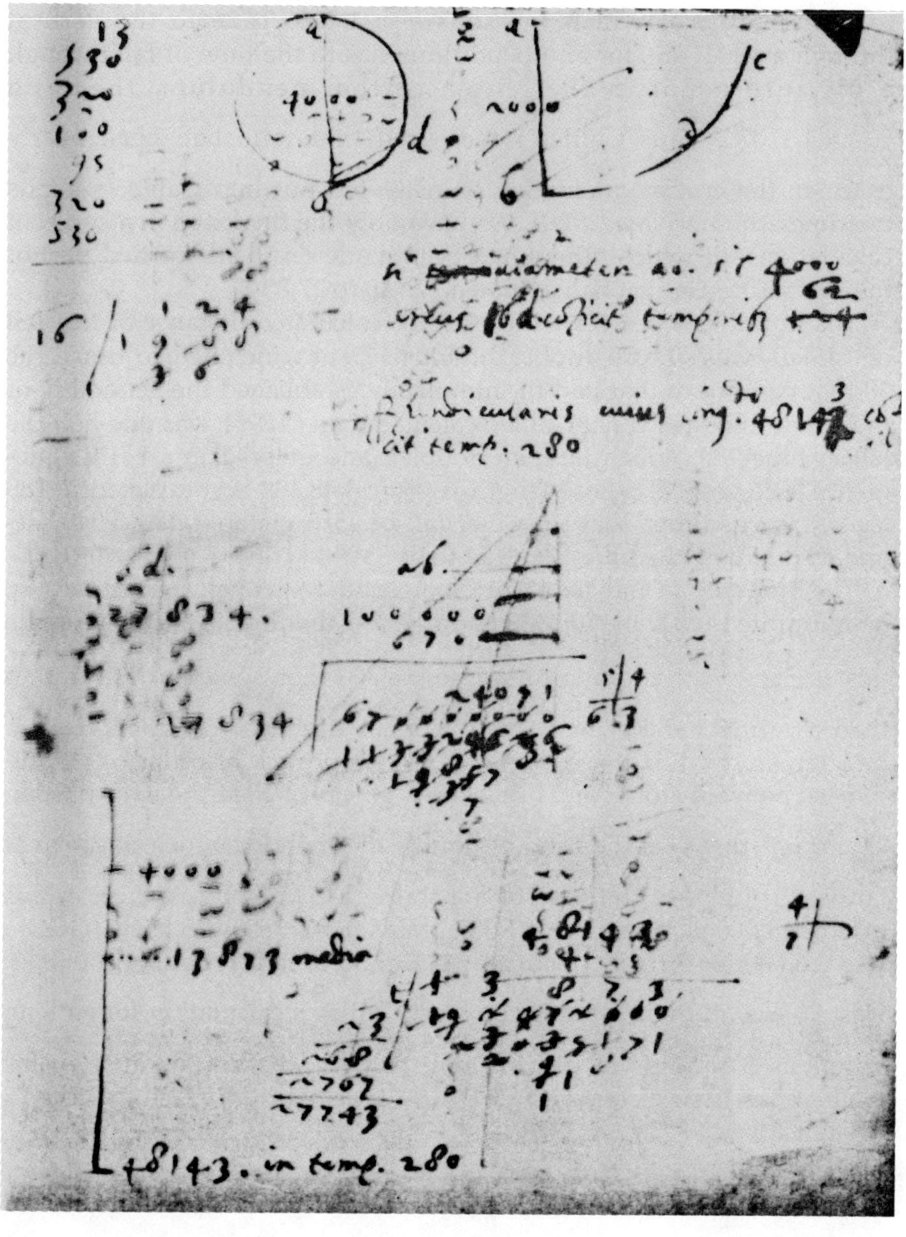

The working paper *f*.189*v*1 on which Galileo was writing when he first recognized the law of fall independently of pendulums. Manoscritti Galileiani in the Biblioteca Nazionale Centrale di Firenze (reproduced by permission).

to four places, of which the square is 1.228. Time to the vertical through a small arc, for any pendulum, has to the time of fall through a distance equal to the length of that pendulum, the ratio $\frac{\pi}{2\sqrt{2}} = 1.1107...$ Thus Galileo's measured time-ratio had been nearly exact. In the crucial calculation on *f*.189*v*1 it put into Galileo's hands the times-squared law of fall. We have only the final step in a series of calculations, of which others had been made on the discarded part of the *f*.90^{b}*v* after Galileo had adjusted 27,840 to 27,834.

The time 280 *tempi,* for which Galileo calculated distance of free fall on *f*.189*v*1, was simply double the 140 *tempi* at which he had timed the 30-foot pendulum that had for him firmly established the generality of the pendulum law. What Galileo calculated on *f*.189*v*1 was one-half the fall in time 280, which he then doubled and entered in a Latin statement: "The vertical whose length is p[unti] 48,143 is completed in 280 *tempi*." In cgs units, that says that fall of 45¼ meters takes 3.043 seconds; I calculate the time at Padua to be 3.038 seconds.

The final step in Galileo's work looks odd; to get half the fall sought, he multiplied 6,700 by 100,000 and divided that by 27,834. If we write this as $\frac{6.7 \times 10^{8}}{27{,}834}$, it is easier to see how 6,700 represented a ratio rather than a number. Because Galileo did not use decimal fractions, but integers (and ratios thereof), he had to keep track of order of magnitude as he went along, and adjust by a power of 10 at the end. Writing $2\frac{T}{t}$ for 6.7, that was the ratio of double the time for which distance of fall was sought, to the time of fall through a known distance, in this instance 4,000 *punti,* timed at 1,337 grains flow = 83.5625 *tempi* (Galileo wrote 6,700 for 6,701½.) Because his timing of fall through 4,000 *punti* was $\frac{1}{30}$ second high, we might expect Galileo's calculation for so long a time as 3 seconds to be very defective. It was not, because in his calculations (using ratios only) t had cancelled out. I factor Galileo's

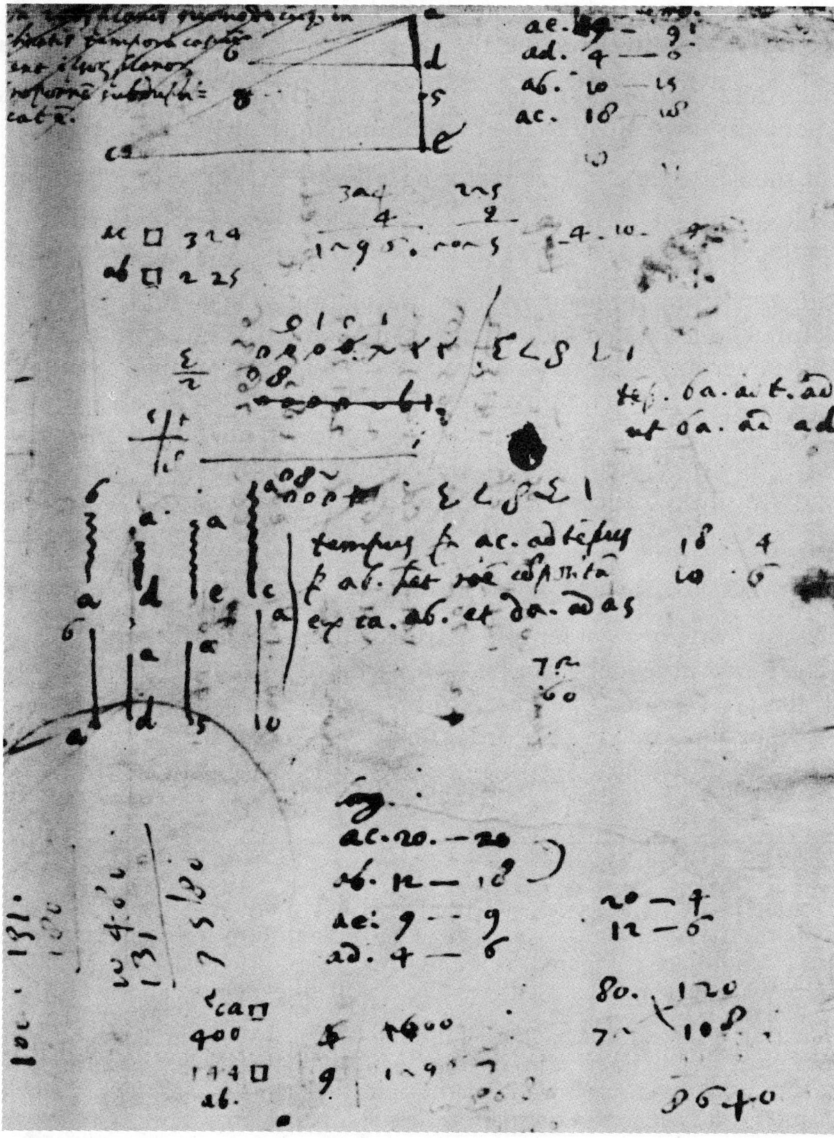

f.189*v*2, with adjustment of Galileo's timing of fall through 4000 *punti* and the general rule for descents along inclines differing in slope and in length. Manoscritti Galileiani in the Biblioteca Nazionale Centrale di Firenze (reproduced by permission).

procedure as below, utilizing the ratio $\frac{f}{p}=\frac{\pi^2}{8}=g$ for the relation that Galileo (had he used symbols) would have thought of as the general ratio of a distance fallen to the pendulum length swinging to the vertical in the same time.[63] This factoring gives: $\left(\frac{Tt}{4}\right)\left(\frac{T}{t}\right)\frac{f}{p}=$ ½ fall in time $T=\left(\frac{f}{p}\right)\left(\frac{T^2}{4}\right)$; and since $\frac{f}{p}=g$, Galileo's end result was mathematically no different from ours using a local gravitational constant.

Before Galileo noted redundancies in his calculations with ratios, he would never have "squared a time" (or a measure of time), an operation that as yet made no physical sense.[64] The product *Tt*, above, arose by compounding *ratios*, a legitimate operation in Euclidean proportion theory, (though Euclid himself used it only geometrically, and did not define "multiply" except for integers.) Nor was the ratio f/p used directly by Galileo in his (now lost) preliminary calculations; it arose as the product of two occurrences of his ratio $\frac{942}{850}$ when calculating from a distance fallen to a length of pendulum, through related times. Galileo's probable steps can be reconstructed in more than one order, with the above clues as a guide to the ratios he formed, though there are further clues making the order fairly certain.[65]

---

[63] The factor that Galileo in fact used was not g but the square root of g, which is the relation between *times* when a pendulum and a fall having the same length are compared. It is the square root of g that I call "Galileo's constant"; the length he used was 1,740 *punti*, as explained above.

[64] Squaring a length did have a legitimate physical sense in terms of areas, and units had long been in use that were analogous to our square foot or square mile. Even now we name no unit the "square second" or "square years," and Galileo did not perform mathematical operations until he perceived them to have a clear meaning.

[65] The principal clues are two uses of a ratio that has not been mentioned, in brief calculations on *f*.189.

Perceiving redundancies in his procedure step by step with ratios, Galileo next took directly the mean proportional between the distances 4,000 and 48,143, recognized this to be almost the half of 27,834, and drew at lower left on *f.*189*v*1 the diagram that he used from that day on to relate times and distances in fall by the mean-proportional rule. On *f.*189*v*2 he used his new-found law to correct his one poor timing (for fall through 4,000 *punti*) from 83½ *tempi* to 80⅔, almost exact. Then on *f.*115*v* he summarized what he had found, this time writing p.(for *punti*) 48,143, confirming my initial assumption that the same unit had been used in 1604 that was first identified from later notes of Galileo's.[66]

Next, having found the law of free fall, Galileo wondered whether it held also for descent along an inclined plane. Taking up again *f.*107*v* (from which we started), he entered in the margin of his tabulation (distances rolled after each of 8 successive equal times) the first eight square numbers. Each, multiplied by the first distance, gave the corresponding measured roll almost exactly. Returning to *f.*189*v*2, Galileo conjectured a compound ratio for descents along planes differing both in slope and in length, found it defective, and by arithmetical test found the correct rule. Using that rule, he calculated on *f.*189*r* his arithmetical confirmation of a theorem he had sent to Guidobaldo del Monte in 1602.

It is no wonder that Galileo preserved all his life *f.*189*v*1, on which he had arrived at the times-squared law of fall. That led him at once to a host of further theorems and solutions of problems, of which the first have just been mentioned. In 1608 he arrived at the parabolic trajectory of horizontally launched projectiles, as the result of a new set of careful measurements of actual uniformly accelerated motions.

---

66 The first page it was possible to explain was *f.*116*v*, belonging to discovery of the parabolic trajectory in 1608. that was done in 1973, leaving discovery of the law of fall still unsolved. In 1975 that appeared to be explained by *f.*107*v*, except for the delayed entry of the square numbers.

6

Applications of the Law

The law of fall remained unpublished until 1632; then it was included (without proof) in Galileo's famous *Dialogue*.[67] Later in that same year, to his annoyance, a related discovery of his was published by Bonaventura Cavalieri, a mathematician of outstanding ability who had studied at Pisa under Galileo's ablest former pupil, Benedetto Castelli. Cavalieri intended no plagiarism; his book was on uses of the parabola, and he included in it a proof of the proposition that when heavy bodies are projected along the horizontal, the paths of fall are parabolic in form. He thought mistakenly that this had already been published, without proof, by Galileo, who was saving it for his final book on motion.[68]

The path of a heavy body falling not from rest, but while it was already moving horizontally, was not identified exactly until four years after the work described in the previous chapter. Its discovery, in 1608, led Galileo to investigate next the case of oblique projection, in which the initial motion is other than horizontal. From his working notes on that problem, much more is now known about Galileo as an experimental physicist than was the case in 1973, when his steps to the parabolic trajectory were first reconstructed.

---

[67] Galileo, *Dialogue Concerning the Two Chief World Systems*, tran. S. Drake (Berkeley, 1953, 1967), cited hereinafter as *Dialogue*. The law of fall appears at p. 222.

[68] When he learned from a friend what Cavalieri was publishing, he was much vexed, but after seeing his *Specchio Ustorio* he spoke of it with admiration in his own *Two New Sciences* (p. 49). This incident shows that the law of fall was put to use by Galileo's acquaintances well before he announced it publicly.

The path taken by a projectile was, for Galileo, a "natural" motion from the moment at which it began falling freely, as when a ball left the hand, or a cannonball left the mouth of a cannon. From that moment on, the motion of the body was subject to the law of fall, and accordingly has its proper place at this stage of the story, though published some years later.

The history of previous speculations about the motion of a heavy body propelled into the air dates from some remarks in the *Physics* of Aristotle. To him, projectiles had a "forced" motion, and precisely for that reason did not merit attention in his own physics as the science of nature. By his definition all "forced" motion was contrary to nature. But near the end of his *Physics*, Aristotle included two views about persistence of "forced" motion after the "force" had ceased (or appeared to have ceased) to act. Some, he said, ascribed this to the air (or other medium) which rushed into the place behind the body to prevent existence of a void space. In so doing, it pressed the body onward. Aristotle disapproved of that explanation, usually called "antiperistasis." He preferred another in which the air acquired some of the force impelling the body before it left the hand (or any agent causing motion.) This moving air, accompanying the body at first, could keep it in motion, though not for long. It was a principle of Aristotle's that anything violent must soon come to an end.

During the Middle Ages the path of a hurled object was more explicitly described. So long as the impelling force remained stronger than the natural tendency to fall, the body continued in a straight line. But that force became diminished by contending against the tendency to fall, which tendency, being natural, was never weakened. Thus a point was reached at which the tendency to fall was more powerful than the supposed remaining force, and the body fell to earth. Belief in a conflict between nature and force obstructed recognition of the independent composition of two motions, even though it had been analyzed in antiquity by the author of *Questions of Mechanics*, then attributed to Aristotle.[69]

In 1537 Tartaglia, as a mathematician, conceived of motion as continuous. He recognized that the path of even the swiftest cannonball must be curved downward from the instant at which it became free to fall, and assumed its two nearly-straight paths to be joined together by a circular arc.[70] In 1546 he emphasized the reasons for his description of the trajectory as not curved at the start, explaining to an artilleryman that for his purpose of analyzing this geometrically, he had treated the "forced" motion as straight, with no great departure from observable facts.

A manuscript by Guidobaldo del Monte contains a remark that trajectories are either parabolic or hyperbolical.[71] He gave the same two methods of drawing their shapes that Galileo offered in his last book, by using a hanging chain or a tilted metal plate on which a rolling ball left its trace as it moved. Perhaps Galileo appropriated these ideas of Guidobaldo, who died in 1607. But the exact reverse may be true, for the two men corresponded and exchanged ideas over many years. The manuscript, now at Paris, is undated, and it is unlikely that Galileo ever saw it.

At the end of his Pisan *De motu*, in 1591, Galileo proposed the ques-

---

69 This, the most ancient treatise on mechanics known, will be mentioned below as Galileo's source for another important idea of his. It was probably written by a pupil of Aristotle's, or by his contemporary Archytas of Tarentum. Galileo owed the basic idea from which he later developed vector-addition to the very first *Question*; cf. *Aristotle, Minor Works* (Loeb edition, 1936) pp. 337:12–339:35.

70 See S. Drake & I.E. Drabkin, *Mechanics in Sixteenth-Century Italy* (Madison, 1969), pp. 84, 103.

71 Cf. Drake & Drabkin, *op. cit.*, p. 48.

tion why forced upward motion remains straight farther, the more nearly vertically projection has begun. His diagram compared several paths differing in steepness of projection, but beyond that he attempted no description of the trajectories. In 1591 Galileo still accepted the concept of "impetus" as a force impressed into the thrown body, regarding its continued motion thereafter as forced. But in 1597, when he lectured on *Questions of Mechanics*, he appears to have perceived from Question 33[72] that an "impressed force" was a superfluous assumption. What remained with a ball could be only the *motion* it had shared with the hand, as Galileo later wrote.[73]

In October 1604, trying to derive the law of fall logically from some incontestable principle,[74] Galileo postulated that all speeds during fall from rest are proportional to the distances traversed. He based that mistaken postulate on his observations and measurements of percussion effects—"machines that act by striking," as he wrote at this time. Those effects are in fact proportional not to the speeds, but to their squares. Galileo had observed that a given weight, falling from a height and then from double the height, strikes twice as hard. He reasoned that the weight being the same, and its speeds different, the speeds must create any difference in effect, and therefore be related as are the distances fallen.

During the next four years Galileo tried various ways of joining this wrong concept of "speed" with his correct theorems about distances and times in fall, always in vain. The reason for his seemingly obtuse, and certainly stubborn, adherence to an incorrect rule of proportionality for speeds was (of course) sound reasoning from an erroneous principle. The principle was that there can be no such thing as mathematically instantaneous speed. Only motion has speed, and motion requires time. The time allowed could be made as small as one wished, but not zero. If the time is supposed to vanish, the motion must vanish along with it. Similar ancient reasoning had afforded the very basis of

---

[72] Question 32 began: "Why do objects thrown ever stop travelling?" and Question 33 went on: "Again, why does the body travel at all except by its self-carriage [αὐτοῦ φοράν] when the discharging force does not follow and continue to push it?"

[73] *Dialogue*, p. 151.

[74] Translated in my *Galileo at Work*, pp. 102–3.

Zeno's paradoxes, to which Aristotle offered many refutations by utterly rejecting the idea of literally instantaneous speed.

Toward the end of 1607 Galileo began arranging his various theorems about motion with the thought of composing a book on the subject. To some proofs still technically uncompleted he added necessary lemmas, and many further theorems occurred to him which he wrote out. In the course of this work he hit upon a solution of the difficulties with regard to "speeds" that had been holding him back. On *f*.164 he drew this diagram, and wrote:

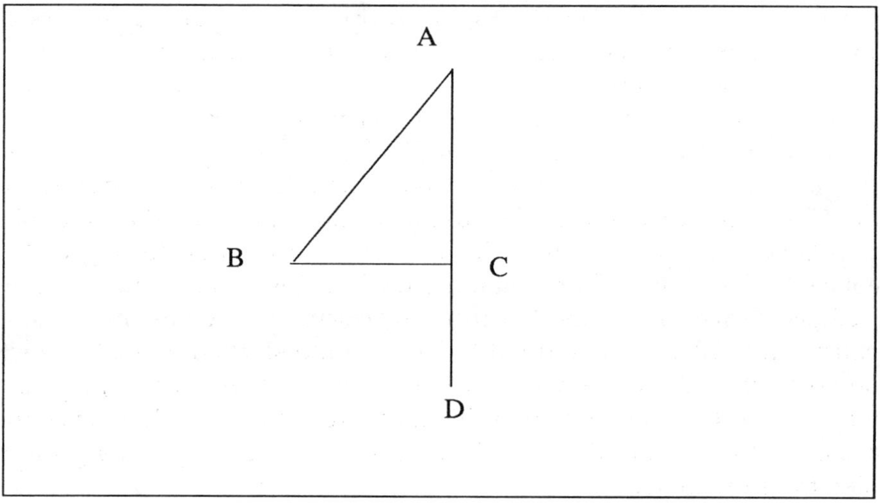

> Marvellous! Now, is motion through the vertical *AD* swifter than that through the incline *AB*? It seems so, since equal spaces are traversed more quickly along *AD* than along *AB*. But it also seems not; because, drawing the horizontal *BC*, the time through *AB* is to the time through *AC* as *AB* is to *AC*, whence the same *momenta* of speed [exist in motion] through *AB* and [through] *AC*; and indeed that speed is one and the same with which, in equal times, unequal spaces are passed that have the same ratio as do the times. The *momenta* of speed of things falling from a height are as the square roots of distances traversed.

Freed from his long misapprehension that speeds in fall were proportional to the distances from rest, Galileo was at last in a position to *measure* speeds in fall correctly. That put into his hands the means of verifying another proposition that he had long believed true, but had had no way of actually testing—that in principle, horizontal motion should continue at uniform speed.[75]

It was early in 1608 that Galileo probably put his inclined plane on a table, let the ball roll down from measured heights, then along the table briefly, and finally fall to the floor. The distance was then measured from each point where the ball struck the floor to the point vertically beneath the end of the table. The speeds of projection being proportional to the square roots of the heights, all horizontal distances measured would follow that same proportion, assuming his longstanding belief correct. His first set of measurements was not preserved, but it probably confirmed Galileo's belief within 4 *punti*; that is, within half a centimeter for projections out to a meter and more.

Some later data are on *f.*116*v* of Galileo's working papers, written probably in the early summer of 1608. When that page was published in 1973, and explained as the discovery-document for the parabolic trajectory, I supposed this to have been Galileo's first testing of the concept I then called "horizontal inertia."[76] Measured distances of projections shown on it departed from his calculations by as much as 40 *punti*, but Galileo appeared not to have been disturbed by that; in fact, he noted the differences with care on the same page. In 1973, nothing whatever was yet known about his measurements in 1604, whence the accuracy to ±3% implied by *f.*116*v* aroused incredulity among other historians of science.

It was generally believed that Galileo had not actually made any experimental measurements at all, let alone accurately. Once Galileo's steps to the pendulum law and the law of fall were fully known, it became evident that *f.*116*v* recorded results of a more sophisticated set of experimental measurements than at first supposed. Notations on that document previously unnoticed, or if noticed, not then under-

---

[75] For his mathematical proof in the Pisan *De motu*, see I.E. Drabkin & S. Drake, Galileo *On Motion and On Mechanics* (Madison 1960), Chapter 14.

[76] Inertia being a dynamic concept, it has no place in Galileo's mature physics. Its place was taken by the conservation of motion as presented in the Second Day of the 1632 *Dialogue*.

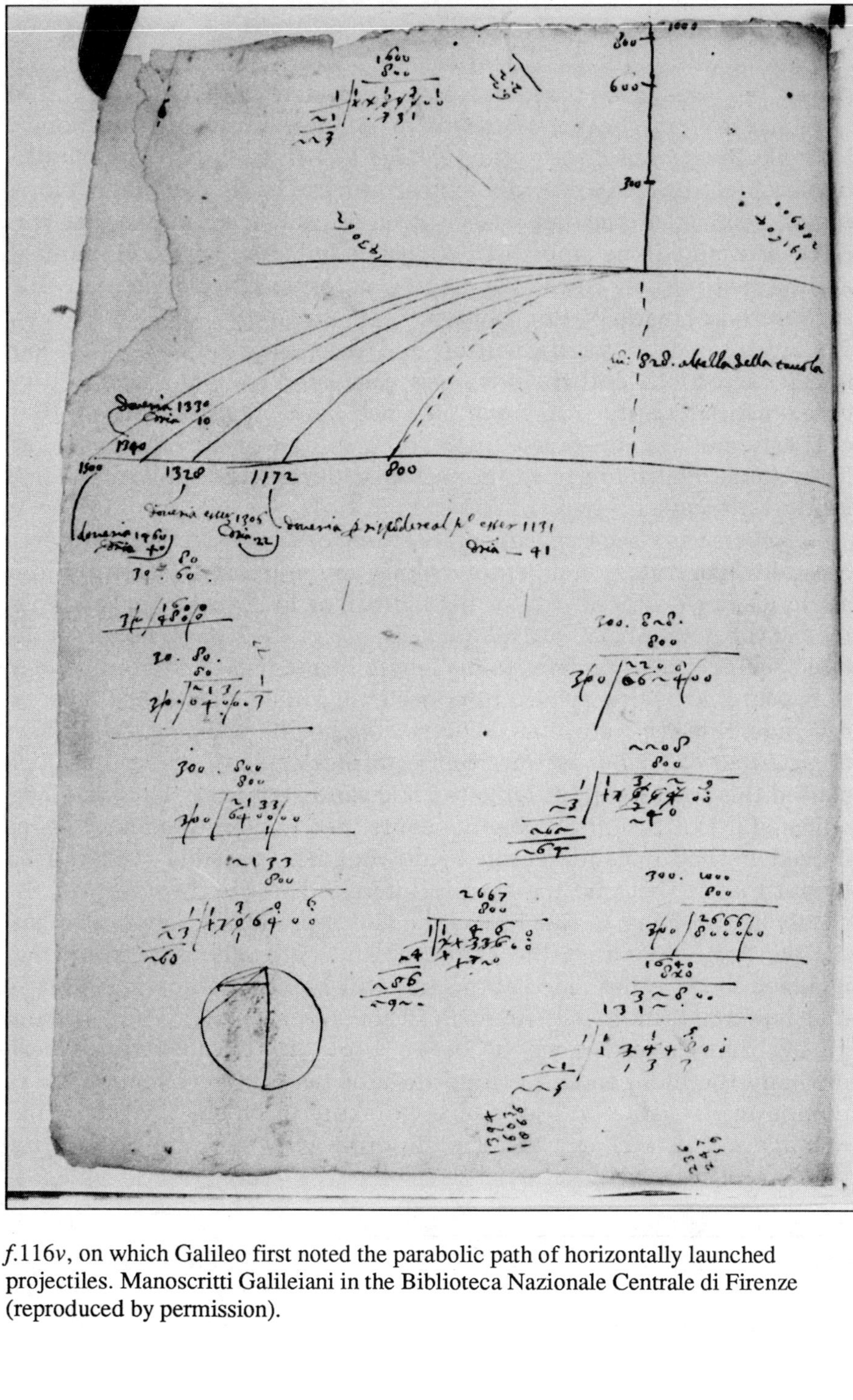

*f.*116*v*, on which Galileo first noted the parabolic path of horizontally launched projectiles. Manoscritti Galileiani in the Biblioteca Nazionale Centrale di Firenze (reproduced by permission).

stood, can now be explained in full.

First of all, Galileo's word *doveria*[77], which I had translated "it should be," was in his time equivalent to modern *dov[e]rebbe*, "it would have had to have been." That was made clear in the 1611 edition of John Florio's *Queen Anne's New World of Words*, an Italian-English dictionary including an appendix on learning the Italian language. Florio remarked that he had never seen even an Italian grammar that correctly explained the fine distinctions in Italian modal conditionals, adding that nothing in the language was responsible for so many absurd mistakes made by Englishmen.

Galileo's reason for measuring and recording the differences between projections and the advances calculated from the first pair of data was that his measures on *f*.116*v* had been obtained in more than one way. He did not *expect* them to be in agreement, and hoped to learn something more from the recorded differences. Nothing of that kind would have been even suspected in 1973.

To reconstruct the experimental set-up for *f*.116*v*, we must observe carefully everything on it before making any unnecessary assumptions about Galileo's procedures, or his figures, or his words. For example, look at the notation: *pū. 828 altezza della tavola*—"828 *punti* height of the table," which clearly relates to the length of a vertical line from table to floor. But it is necessary also to notice that, unlike most of the lines on the page, this is a *dotted* line. Whether the height of the table was ever adjusted to 828 *punti* is uncertain; but entries show us that Galileo started this work with his table-top 800 *punti* above the floor, and later adjusted it to 820 *punti*. Probably he intended to raise it to 828 *punti*, as indeed he had a valid reason to do. But it is possible (and indeed likely) that he became too deeply interested in his discovery of the parabolic trajectory to finish what he had been doing. For Galileo had not started out to find the shape of any path; he came upon that incidentally to other calculations, as will be seen. Knowing already that horizontal motion is uniform in the absence of any impediment greater than air resistance, he began work on *f*.116 to verify experimentally the independent composition of two different tendencies to motion in the same body, a conception sure to be challenged by upholders of Aristotelian natural philosophy who were by 1608 in the

---

[77] doveria = dovrebbe

habit of automatically contradicting everything that Galileo wrote.

At the top center of *f.116v*, Galileo calculated the mean-proportional of a number and its double, in his usual manner. We would now multiply the smaller number by $\sqrt{2}$, but Galileo took the square root of their product. The first number was 800. Most of the other calculations—which involved more than a number and a double—are less simple. But at lower right, and unused in the body of the document, the short-cut calculation was repeated for 820. It was not repeated for 828, which would have given 1,171.

There is more than one way to carry out these measurements, and they do not all give the same results. Probably Galileo had already simply set his grooved inclined plane on a table, without any concern about either its height or the angle of the plane. That was done early in 1608, and his measurements were not saved. Coming off the plane, the ball would strike the table and bounce. That would not greatly matter, so long as it rolled on the table before it fell to the floor.[78] Galileo did not know whether this made a difference, but recognized that it might.

To eliminate or at least reduce any bouncing, two procedures were available. A board of the precise thickness to receive the ball on its lowest point exactly when it left the plane could be placed on the table, or a curved deflector could be grooved to match the width of guiding groove in the plane. Those procedures will not yield identical lengths of projection. The first causes the ball to roll on its lowest point, at least briefly; and that will suffice to increase its speed just before it starts to fall freely. While in the groove, the ball rolls on a reduced radius. If it is then allowed to roll on its entire radius, some of its energy of rotation is converted into speed of forward advance. By changing the mode of deflection, Galileo had increased the length of his initial projection from 800+ *punti* to 820 *punti*.

No doubt he was surprised by this, because he had meant only to eliminate, or reduce, any bouncing. But having seen that the distance of projection was also affected, it naturally left him wondering how *much* it could be increased. Before describing how he found out, I shall first explain the +, added above after 800. I calculate projection to be

---

78 Very little energy would be lost by the bounce of a bronze ball on a hardwood table. The main thing was to assure horizontal motion at the time the ball left the table, as will be seen.

804—about where the *solid* curved line in Galileo's diagram meets the floor. He then drew a *dotted* line to represent 800, a projection not actually measured, but it was much simpler to use in calculations of mean proportionals by the extraction of square roots.

A guiding groove is quite necessary when the plane is given a considerable tilt, to assure projection always along the same line. Even a small imperfection in the groove will send the ball flying out of it when the motion becomes very swift. To prevent that, Galileo covered the groove with limp vellum; the weight of the bronze ball pressed this down enough to keep it rolling quite straight.[79] Thus Galileo noted 828 *punti* to be the maximum initial projection, after vertical descent of 300 *punti* along the plane. It is very doubtful that he ever knew the reason for this, which is that the lowest point of the ball remained in contact with the vellum covering during its entire roll.

Next comes the question why 828 *punti* should appear as the height of the table, when it had been found as the maximum length of projection for descent of 300 *punti*. There was a reason for this. Vertical descent during roll along the plane and height of the final drop to the floor are interchangeable in this kind of experimental measurements. Distance of projection remains the same for vertical descent x and drop y as for vertical descent y and drop x. As said earlier, whether Galileo completed such a trial is uncertain, but there can be no reasonable doubt that he had discovered experimentally a great deal more about motions on inclined planes (and off them) than he eventually published.

His table was about the normal height, and the highest that Galileo had to reach when placing the ball was about six feet, so he could easily bend down to observe its paths. Those were then sketched on *f.*116*v*, and they appear parabolic. The calculations on the page are such as to have assured Galileo that they must be indeed semi-parabolas. A letter that he wrote in February, 1609, confirms that he was then studying phenomena of artillery shots, and a diagram in it also contains a sketch of a parabolic path.

---

[79] Cf. *Two New Sciences*, p. 169.

As mentioned at the beginning, all previous writers on the paths of projectiles regarded those motions not as natural, but "forced"—including Galileo himself in his youth. But after he abandoned the concept of "impressed force" in favor of conserved motion, he came to regard the motions of projectiles as natural motions from the moment when any outside force ceased to act. In 1609 he started making tables of heights and distances for shots fired at different elevations of a mortar, later completed and published in his last book.[80] The symmetry of parabolas makes it possible to reduce such calculations to equivalent horizontal trajectories. But there is another class of oblique projections that are not so reducible, and to that Galileo next turned his attention.

When a ball rolls from the end of an inclined plane without any horizontal deflection, its fall to the floor is not a *simple* parabolic curve, though this is a *natural* motion from the moment that a finger is lifted from a ball held at rest on an incline. No force is imparted to the ball at any time (Galileo did not regard gravity as a *force,* but as a *natural* tendency downward.) These constituted a new class of projections belonging to his physics as the science of natural motions, and on *f.*114*v* there is a freehand sketch of seven descents, after rolls of different lengths along an inclined plane, with the projections measured in *punti*. Also, on *f.*81*r,* we have a diagram on which were recorded the results of some quite sophisticated (and remarkably accurate) experimental measurements of oblique projections. From those records it is possible to determine exactly what equipment was used by Galileo, and how he went about investigating a problem in physics that he was unable to solve, and about which he therefore never published anything.

In 1973, calculating from the data on *f.*116*v* (using modern information and procedures), I noticed that there was an apparent systematic loss of gravitational energy in Galileo's results. At that time I employed this, incorrectly, to estimate the angle at which the plane had been tilted. That angle does not matter when all projections are horizontal, as critics were quick to point out. Eventually I realized that the

---

[80] There Galileo took care to point out that his tables would not apply to high-speed artillery shots, because resistance of the air became great enough to alter the shape of trajectory very considerably.

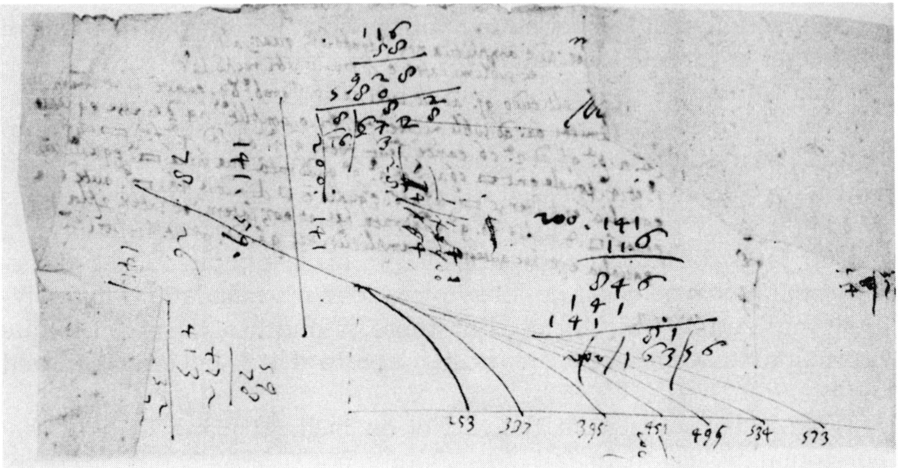

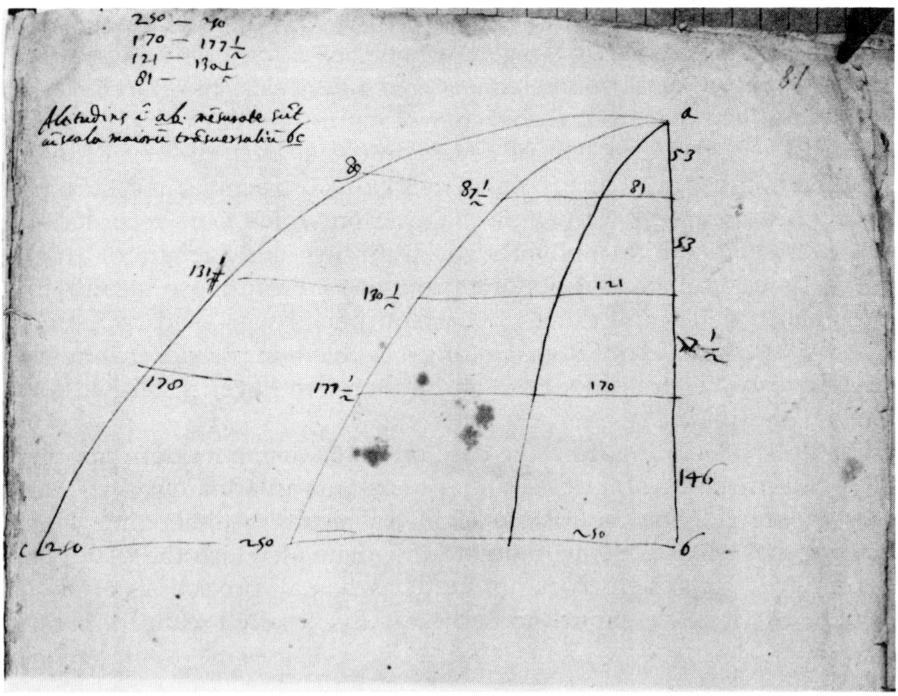

Galileo's records of measured projections from the ends of inclined planes without horizontal deflection.
Above, part of *f.*114*v*; and below, *f.*81*r*. Manoscritti Galileiani in the Biblioteca Nazionale Centrale di Firenze (reproduced by permission).

seeming loss was the result of his using a grooved plane, into which part of the ball always extends. Its forward acceleration is thus diminished, while the equivalent energy goes into rotating the ball. No energy, of course, is lost, and Galileo's measurements proved to have been uncannily accurate when all the physical factors were taken into account.

It is from the data concerning oblique trajectories of the kind described above that the diameter of Galileo's bronze ball and the width of the guiding groove in his plane can be deduced. The ball was 20 *punti,* or just under 19 mm, in diameter, and the groove was between 8mm and 9mm wide. The same plane was used in 1604 and 1608–9, being a little more than 2 meters long. In his final work on oblique trajectories Galileo added another plane, described in his last book as 12 *braccia* (or about 7 meters) in length. This was butted against the other plane to permit a roll of 9,600 *punti,* required for the work recorded on *f*.81*r*.

Also from the data on oblique projections it is possible to deduce the angles at which Galileo fixed his plane, and why he chose those angles. The freehand sketch on *f*.114*v* indicates an angle of about 26° (as I noted in 1973), and the data on *f*.81*r* later threw light this seemingly strange choice. More precisely, Galileo's angle was 26.57°, which is arctan ½, an angle that can be constructed very exactly by two simple measurements. One has only to make sure that the height above the table is exactly one-half the distance along it from one end of the plane to the other. Galileo's reason for that choice was that he wanted the horizontal advance at the moment of projection to be double the vertical motion downward.

There is abundant evidence that the work on *f*.116*v* was also carried out with the plane at this same angle, once all the data in the working papers are fully understood. For the measurements behind *f*.81*r*, three different angles of plane were employed, and three lengths of roll, and three heights of drop. The angles chosen were arcsin ⅓, arcsin ⅙, and arcsin 1⁄12, or 19.47°, 9.59°, and 4.78°—all seemingly irregular, but actually easy to set with precision and designed to maximize Galileo's chances of recognizing simple numerical ratios, if any were to be found.[81]

These investigations of the paths of bodies in free fall after motion initiated by descent along inclined planes serve to illustrate the fruitfulness to mathematical physics of the law of free fall. In the course of Galileo's applying it, he hit upon a new idea for explaining the places and speeds of the planets in terms of the law of fall, outlined many years later in his famous *Dialogue* of 1632.[82] In a very different way, Johann Kepler[83] had attempted to account for the planetary orbits in his first book, from which Galileo had taken the data that had led him to his discovery in 1601 that planetary distances from the sun could be related directly to orbital speeds in the Copernican system. One entry added in 1609 to an earlier diagram of concentric circles, representing planets with the sun at the center, cannot have been made before the writing of *f*.91, on which Galileo proved that times in fall from rest are measures of speeds acquired. Close similarity between the form of his discovery in 1601 and that of the law of fall then led Galileo to the "Platonic cosmogony" that he set forth near the beginning of his 1632 *Dialogue*. There he speculated that all planets were created at a place beyond Saturn (then the outermost planet known), and moved toward the sun with uniformly accelerated motion until each had reached the speed at which God had ordained it to circle the sun forever. Planetary

---

81 The first two angles and rolls were such as to produce a speed of projection and its double. For the third, that would have required an impracticably long plane; Galileo used half that length, and even that required him to make a second plane about 20 feet long.

82 *Dialogue*, p. 29.

83 See the Epilogue for more on "Kepler's problem," as this may be called.

deflection from straight accelerated to uniform circular motion was by special act of the divine will, in Galileo's scheme.

This speculation is wrong, but not randomly so; it is wrong by a factor of 2, so to speak.[84] Newton, asked about it, replied that for the sun to *hold* the planets in orbit in such a case, its gravitational power would have to be doubled at the instant of deflection into circular motion. Galileo certainly had no idea whatever of universal gravitation, and in the *Dialogue* he denied that he, or anyone, knew what held planets in orbit.[85] Yet it is interesting that his mathematics of ratios and proportionality, applied to physical concepts such as distance, speed, and time, led him to a speculation that is not entirely unrelated to facts undiscovered until long after his death.

It is noteworthy that Galileo put the times-squared law of distances in spontaneous natural descent to use, applying it to previously unsolved problems of physics, and so did Cavalieri, independently, arriving at one of the same solutions. During a whole decade following its publication in 1638 by Galileo and also by a friend and correspondent of his, nearly all professors of natural philosophy rejected it on metaphysical grounds of the kinds summarized in the next chapter. The two views of science, one as something useful to mankind and the other as only a branch of philosophy, still live on in wordy disputes today.

---

[84] Because Galileo worked entirely with ratios, the factor of 2 in both terms would simply have cancelled out. Those who scorn his "Platonic cosmogony" as unscientific do not take into account the mathematics he employed when they resort to algebraic equations. What those prove is that the scheme was wrong, not unscientific. Most scientific theories started out in incorrect forms.

[85] *Dialogue*, p. 234.

7

Reception of the Law, 1632–49

Galileo first publicly announced his law of fall in 1632, using it in his *Dialogue* for calculating time of fall from the moon to the earth. Only a "probable argument" in support of one consequence[86] of the law was offered in 1632. Not until 1638, in his *Two New Sciences,* was the law derived from the definition of uniformly accelerated motion as motion in which equal increments of speed are acquired in equal intervals of time. It is highly probable that Galileo was still unable to formulate a rigorous proof of the law from that definition alone until shortly before his last book went to press, as one scholium[87] in it appears to allude to a proof resembling the "probable argument" set forth in the earlier *Dialogue,* not used in the printed text of 1638.

The occasion for stating the law of fall in the *Dialogue* was an absurd calculation by the Jesuit Christopher Scheiner[88] of the time it would take a cannonball to fall to earth from the moon;. Assuming the fall to

---

86 The double-distance rule for uniform motion following uniformly accelerated motion from rest, which Heytesbury had deduced in the 14th century. See Galileo, *Dialogue* (tr. S. Drake, Berkeley 1953), pp. 227–30.

87 To Proposition 23; see Galileo, *Two New Sciences* (tr. S. Drake, 2d. ed., Toronto 1989), p. 196. When this scholium was written, proof of Proposition 1 probably still depended on an appeal to areas. In its final form (pp. 165–6) the appeal is to one-to-one correspondence of lines and congruence of triangles which contain them.

88 At the time Scheiner published his statement he was not unfriendly toward Galileo, but by the time of the *Dialogue* he had become a vindictive enemy over priority in the discovery of sunspots.

be straight to earth at 12,600 German miles per hour, taken as the orbital speed, Scheiner put the time at more than six days. On the author's assumptions, Galileo noted, the time should be less than four hours, the radius of orbit being less than one-sixth of the circumference described in 24 hours, making his claim wrong by a factor of 36. For a correct calculation, Galileo's spokesman said:

> First of all, it is necessary to reflect that the motion of descending bodies is not uniform, but that starting from rest they are continually accelerated... But this general knowledge is of no value unless one knows the ratio according to which the increase takes place...

The odd-number rule of distances was next stated, and then the law of fall:

> ...In sum, this is the same as to say that the spaces passed over by the body starting from rest have to each other the ratio of the squares of the times in which those spaces were traversed. Or we may say that the spaces passed over are to each other as the squares of the times.[89]

This statement of the law went unnoticed among the hostile critics of the *Dialogue*,[90] perhaps because it did not explicitly mention the proportionality of speeds in fall directly to times, which might have aroused some adherent of Albert of Saxony, or a follower of Varro, to object. Friendly critics, however, noted that in preparing to make his own calculation, Galileo wrote:

> ...let us suppose we want to make the computations for an iron [cannon]ball of 100 pounds which in repeated experiments falls from a height of 100 *braccia* in 5 seconds...

---

89 *Dialogue*, pp. 221–2.

90 Pierre Fermat was anything but a hostile critic, praising the book highly when he sent it to Pierre Carcavy. But Fermat thought he had proved the law to be false. After Galileo's derivation of it in *Two New Sciences* appeared and became the subject of heated debates, Fermat supported the law; see below.

Clearly, round figures were taken here in order to make the ensuing calculation simple, and easy for readers to follow. But Marin Mersenne took this to be an assertion of fact. After he verified the length of the Florentine *braccio,* he felt certain that Galileo had never carried out careful experiments. From Genoa, G. B. Baliani wrote to Galileo asking how he had arrived at his figures, which did not conform with Baliani's findings. And long afterward, in 1665–6, Newton took the above passage as describing actual results, although it is obvious that neither Galileo, nor anyone else, had made repeated experiments with a 100-pound cannonball (if there was such a thing) from a height of nearly 200 feet.[91]

Baliani, who had been in correspondence with Galileo since 1614 and visited him at Florence in 1615, was in 1611 in charge of the arsenal at Savona, where he had first made experiments relating to the fall of heavy bodies. It was doubtless in 1615 that Galileo gave him the information of which Baliani wrote to Benedetto Castelli in 1627:[92]

> ...Composing the treatise on solids of which I have spoken, it happened that, without seeking this, I found (as it appears to me) well proved in a very surprising way a proposition that Signor Galileo had told me to be true without his adducing to me the demonstration; and this is that in natural motion, bodies go increasing the speeds in the proportion of 1, 3, 5, 7, etc., indefinitely; however, he adduced to me a probable reason, that only in this proportion greater, or lesser, spaces preserve always the same proportion. I do not explain further because I know I am speaking with one who understands...

---

91 It is a commentary on the state of the literature in 1665 that Newton did not have access to authentic experimental measurements of acceleration in fall, three decades after the law became known. Galileo eventually replied to Baliani that for the purpose of showing the enormous error of his opponent, it made no difference what numbers were used in this calculation.

92 Antonio Favaro, ed., *Opere di Galileo...* (Florence, 1935) vol. 13, p. 248.

The treatise on solids to which Baliani alluded was the book on motions of heavy bodies he published in 1638, deriving the law of fall from the pendulum law, which he postulated from actual measurements. He sent a copy to Galileo,[93] who had become blind and could not study the diagrams accompanying Baliani's theorems. Thanking Baliani for the book, Galileo remarked that in his own book he had reasoned *ex suppositione,* in the style of Archimedes. He had then offered experimental evidence that his definition of uniform acceleration agreed with natural phenomena, and in that agreement, Galileo told Baliani, he counted himself fortunate. In the same letter he explained why he had used arbitrary data in the *Dialogue,*[94] and how a correct determination could be made by using a very long pendulum and counting its oscillations during two successive crossings of the meridian by a fixed star.

It is noteworthy that when the times-squared law of fall was eventually found and proved in the 17th century, the independent experiments of Galileo and Baliani that led to it had led also to the law of the pendulum. That law was of interest to musicians; it appears to have been first discussed among them in the 1630's, when quite possibly it had been independently discovered by Marin Mersenne in France or G. B. Doni in Italy, and perhaps both. But pendulums had played no part in any theory of fall before the 17th century, and seemed not to interest the natural philosophers who opposed the law of fall in the years following 1638.

The best-known supporters of the fall law were Mersenne and Pierre Gassendi, both at Paris. Its two most vigorous opponents were also French—Pierre Cazré and Honoré Fabri.[95] Cazré cited impact ex-

---

[93] G.B. Baliani, *De motu naturali gravium solidorum* (Genoa, 1638). Baliani was unaware that *Two New Sciences* had been published, and though some copies had reached Rome, Galileo had only a proof copy sent to him for preparation of errata.

[94] Galileo's troubles with the Roman inquisition over that book had caused him to neglect replying to Baliani's earlier inquiry. The letter is translated in my *Galileo at Work* (Chicago 1978), pp. 399–401.

[95] Whether Baliani should be called an opponent of the law is a complicated question. In 1638 he offered his own proof in its support, but in 1646 he proposed a different law, with an analysis very similar to that of Fabri in the same year.

periments in support of a quite different law of fall, while Fabri conceded that Galileo's experiments implied the times-squared law and simply denied that measurements of any kind, however precise, could reveal the philosophically true laws of nature—for fall or anything else. Curiously enough, the only mathematician outside Italy who thoroughly understood Galileo's derivation of the times-squared law was also French. This was Jacques-Alexandre Le Tenneur, who vindicated Galileo's law in 1649, writing at Mersenne's request after Fabri's publication. Like his contemporary, the celebrated mathematician Pierre Fermat, Le Tenneur was a lawyer and counsellor to a provincial parliament.

Mersenne, an outstanding authority on musical history and theory, was an ingenious and careful experimentalist in physics whose correspondence with scientists and philosophers of the 17th century is the source of much information about the law of fall. In particular, the thought of René Descartes on this subject is known only from his letters to Mersenne. Descartes' interest in the law of fall began with a question addressed to him by Isaac Beeckman in 1618. Beeckman's own answer to this, not the same as that of Descartes, was entered in his journal for that year. It is of interest as reflecting a physical approach similar to that of Albert of Saxony (as I understand that) when he mathematicized impetus theory in the 14th century, except that attraction by the earth now took the place of impetus. Beeckman wrote:[96]

> ...With this agreed, things are thus moved downward toward the center of the earth [through] empty space. In the first moment, as much space is traversed as possible by attraction of the earth. With this motion continuing, in the second moment a new motion is added by attraction, so that double the space is run through in this second moment. In the third moment the doubled space remains and to it is added a third by attraction of the earth, so that in [this] one moment a space triple the first space is run through...

In Beeckman's mathematization, this separation of numbered moments was then abandoned, and continuous uniformly accelerated motion emerged—if we allow the kind of reasoning by which one could also prove that the diagonal of a square equals the sum of two

---

[96] In his journal for late 1618; this and later passages are from Clagett, pp. 417–18.

sides of it. In the opening lecture of a course in calculus it is not unusual to offer this startling proposition initially, drawing a right triangle whose hypotenuse is not a line but a series of small steps. Students are then invited to consider the steps more numerous, and smaller, until they become too small to imagine, and are asked whether they could then safely treat the result as a line. In Beeckman's diagram, *ADE* and *ACB* were equal right triangles, so placed that *A [E] B* was the diagonal of a square whose side was double *AD*. His proposition was:

> ...Since these moments are *individua,* we have a space such as *ADE* through which a thing falls in one hour. The space through which it falls in two hours doubles the time-ratio [i.e., is to the space fallen in the first hour as the square of the time-ratio]; that is, the ratio of *ADE* to *ACB* is the square of the ratio of *AD* to *AC*.

The argument proceeded by dividing each square into smaller squares and noting that those may be again divided interminably. No appeal whatever was made to the rigorous Euclidean theory of proportionality for mathematically continuous magnitudes. The vanishingly small areas were simply regarded as points, in this reasoning, which was not essentially different from that used by G.P. de Roberval in his treatise on indivisibles[97] two decades later.

Beeckman reasoned only about motions, and though he spoke of attraction toward the earth, he did not introduce the idea of any force. His account of fall was purely kinematic—and although the validity of his mathematical argument is dubious, he thus arrived at the times-squared relationship. Descartes understood his question differently and dealt it as a problem in which force was to be considered. He reasoned as if a whole square could not be ignored just because its area is small, though it would be all right to disregard half of it. His dynamic solution was not the times-squared law. Nevertheless, after reading *Two New Sciences* twenty years later, Descartes wrote to Mersenne that he had once arrived at the same conclusion as Galileo,

---

97 It was this that induced Newton to remark that the hypothesis of indivisibles seems a trifle harsh. Roberval's use of the word "indivisibles" was rooted in medieval denunciations of them, in turn traceable to the pseudo-Aristotelian *On atomic lines*. In 1635 Bonaventura Cavalieri used the same word very differently, to denote elements having one less dimension than the magnitude treated as containing them.

but then came to see that there is no unique law of fall. Every body falls in the way best suited to its material essence, which could be learned only by studying the Cartesian principles of philosophy. It might happen that something fell according to Galileo's law, but that was not usually the case in Descartes' mature opinion.

In 1639 Mersenne published in French an abridged paraphrase of *Two New Sciences*. Though he approved Galileo's law of fall, he added that he saw no reason for rejecting proportionality of speeds in fall to distances fallen. Mersenne's prowess as an experimentalist was not matched by mathematical perceptivity. In 1645, at Paris, a book was published by Pierre Cazré, S.J., which bore the scornfully anti-Galilean title:

> Demonstrative Physics, in which are determined the ratio, measure, mode, and power of acceleration of motion in natural descent of heavy bodies, against the pseudo-science of the same motion recently thought up by Galileo Galilei, Florentine philosopher and mathematician.

Pierre Gassendi, after corresponding with the Jesuit father, replied publicly in defense of Galileo in 1646, also at Paris, with a book entitled:

> On the proportion in which falling heavy bodies are accelerated. Three letters. In reply to as many letters by the reverend father Pierre Cazré, S.J.

One principal point at issue among natural philosophers and mathematicians during 1638–49 was to be Galileo's assertion, in *Two New Sciences*, that it was as false and impossible for uniform acceleration to be proportional to distances fallen (or to be fallen) from rest, as it is for instantaneous motion to occur. That statement, which had been at least implicitly questioned by Mersenne, was frequently challenged in the 1640s. It continues to be branded as false by some. A number of modern critics of Galileo, beginning with Ernst Mach, have made a point of refuting it by appealing to a simple differential equation.[98] Galileo's statement had been made in reply to this assertion by Sagredo:[99]

> That the falling heavy body acquires force in going, the speed increasing

---

[98] Of course it is possible to differentiate with respect to distance instead of time. The result is an exponential equation in which the body cannot start from rest, a concept satisfactory to mathematicians but a trifle harsh for physicists.

[99] *Two New Sciences*, p. 160.

> in the ratio of the space, while the momentum of any given percussent is double when it comes from double height, appear to me as propositions to be granted without repugnance or controversy.

To which Salviati replied, in part, as follows:

> And yet they are as false and impossible as that motion should be made instantaneously, and here is a very clear proof of it. When speeds have the same ratio as the spaces passed or to be passed, those spaces come to be passed in equal times; if therefore the speeds with which the falling body passed the space of four *braccia* were the doubles of the speeds with which it passed the first two *braccia*, as one space is double the other space, then the times of those passages are equal; but for the same movable body to pass the four *braccia* and the two in the same time cannot take place except in instantaneous motion.

There was more to the reply, but this completed his "clear proof." Cazré's opposition related chiefly to the ensuing part, concerning the force of impact. Gassendi replied at great length, exposing many fallacies in Cazré's alternative law of fall, which in effect made speeds through successive equal distances increase as successive integral powers of 2. Cazré's impact experiments, interpreted to refute Galileo's law of fall, made his theory a dynamic one.

About a year before he discovered the law of fall, Galileo had carried out equivalent experiments. Later, misled by their results,[100] he attempted to derive the times-squared law of fall from the "effects of machines that work by striking," which led to his famous letter to Paolo Sarpi of 16 October 1604. But the matter of present interest is reception of the times-squared law after its proof, not the evolution of that proof.[101]

Salviati's reply to Sagredo was explicitly called a clear proof of the falsity and impossibility of two propositions, and that is what it was.

---

[100] He took the impact effect as a measure of speed, whereas in fact that depends on the square of the speed acquired in fall. From 1608 on he avoided this mistake, into which Cazré fell later.

[101] The relation of Galileo's first attempted derivation of the law of fall, written for Sarpi, to the definition of "speed" at which he arrived in 1608, was discussed in my "Galileo's discovery of the law of free fall," *Scientific American* (June 1973), pp. 84–92. At that time I believed the discovery to have involved a bit of good luck.

Yet even the peerless mathematician Fermat did not recognize it as having *proved* anything. Writing to Gassendi in 1646, Fermat said that doubtless Galileo had had a proof, but did not set it forth. Rather than wasting valuable time in long replies to Jesuit opponents, he said, the task was to supply a proper mathematical proof for Galileo. This Fermat proceeded to do, in classical Archimedean style, and his proof took a half-dozen pages.[102] There is no getting around Fermat's argument in support of Galileo's position, and there is also no way of substantially shortening that argument that I can find. But neither was there anything wrong with Galileo's "very clear proof" in 1638, which took only one sentence. Fermat simply did not recognize it to be a mathematical proof because it assumed a principle banned from classical Greek mathematics by the final axiom in Euclid's *Elements,* Book I,—"The whole is greater than the part."

That axiom excluded the infinite from Euclidean mathematics, perhaps because Greek philosophers had got themselves into a host of paradoxes by attempting to deal with it. Galileo introduced the infinite into Renaissance mathematics in the First Day of *Two New Sciences* by applying the concept of one-to-one correspondence between members of two infinite sets to the positive integers and their squares.[103] In the Third Day that same concept, applied to geometry, gave him his basis for deriving the law of fall. In 1646 Baliani published the second edition of his 1638 book on the motions of heavy bodies and of fluids, with an added section asserting that the odd-number rule of spaces in equal successive times during fall was not exactly true. In that new section, Baliani's original kinematic account of fall remained as before, but the new section gave a dynamic account, related to medieval impetus theory but offering a mathematical analysis differing completely from that of Albert of Saxony. Baliani now declared that the times-squared law was only an approximation to the truth.

The year 1646 saw not only Gassendi's reply to Cazré and Baliani's

---

[102] *Oeuvres de Pierre Fermat,* vol. III (Paris, 1896).

[103] See pp. 40–44 in the translation cited earlier. The discussion was preceded and followed by geometrical and material examples.

shift from a kinematic to a dynamic theory, but a third book, Fabri's detailed treatment of impetus theory in opposition to Galileo, published at Lyons by his pupil Mousnier.[104] The last- named book gave occasion to Mersenne, whose specialty was not mathematical analysis, to invite Le Tenneur's comments, resulting in the first (and for a long time the only) book[105] whose author clearly understood the principle of one-to-one correspondence and saw its role in Galileo's one-sentence argument in refutation of the hypothesis of proportionality of speeds to distances in fall from rest. One paragraph from Le Tenneur's book cannot do it justice, but will show how briefly the proportionality of speeds in fall to distances fallen could be refuted mathematically:

> If possible let the heavy body fall through two equal spaces *AB* and *BC* so that the speed at *C* has become double that which it had at *B*. Certainly, under the hypothesis, there is no point in the line *AC* at which the speed is not double that at the [unique] homologous point in line *AB*... Therefore the speed through all *AC* was double the speed through all *AB*, just as the space *AC* is double the space *AB*; therefore *AC* and *AB* are traversed in equal time.

This, of course, comes down to saying with Galileo that the hypothesis is as false and impossible as is instantaneous motion. For speeds to be as distances fallen, some discontinuity must exist in motion from rest. A body cannot fall through distances *d* and 2*d* in the same time in *continuously* accelerated motion, in the mathematical sense of "continuous."

The impetus-theory of fall by quantum-jumps, from rest to a uniform speed and from that to the next, entailed an infinitude of discontinuities in speed. That had not disturbed Buridan or Albert of Saxony, who regarded motion as a successive process and treated accelerated motion as proceeding by a succession of new speeds.[106] The mathematics of continuity was not yet accessible at their time because

---

104 *Tractatus physicus de motu locali* (Lyons, 1646).

105 J.A. Le Tenneur, *De motu naturaliter accelerato* (Paris, 1649).

106 Discontinuity occurred at the end of each part of the motion, in Albert's analysis as in Buridan's description. Only the "first" motion from rest was free of impetus. The initial discontinuity at its end being allowed, all the others followed logically and causal explanation was complete for any number of parts taken.

of the medieval corruption of Euclid, Book V.

Even after Galileo applied the mathematics of continuity to acceleration in fall, discontinuity did not bother Fabri, perhaps because the propriety of regarding very small things as simply non-existent—valuable in philosophy, but hazardous in strict mathematics—went unquestioned among natural philosophers who demanded causes above all else in scientific reasoning.

Like modern quantum theory, the impetus theory of fall entailed an indeterminacy, Each uniform speed, however brief, corresponded with neither a unique point in space nor a unique instant of time. Natural philosophers did not see indeterminacy as an unsatisfactory element in causal accounts of phenomena. Indeterminacy had been given a name in the Middle Ages—"the latitude of forms"—which thereby became an entity deserving analysis, in the scholastic tradition of terminological debates.

What is surprising, in the 17th-century rebirth of impetus theory to combat the times-squared law, is not merely the revival of discontinuity and indeterminacy in physics, but also the clear inability of even some mathematicians, as natural philosophers, to recognize lacunae in their own reasoning after the rationale of continuity had been reintroduced by restoring Euclid, Book V. Fabri, who has to his credit some valuable contributions to the beginnings of modern mathematical physics, is an example of this surprising loyalty to traditional physics.

Fabri wrote, "A body falling freely will go faster in the vertical than it will along an inclined plane. Therefore it must start faster. Hence there must be some speeds along the plane that are not present in straight fall." Galileo had said that in order to reach any speed from rest, a body must pass through every lesser speed. To postulate actual infinitude of speeds was an inadmissible assumption for Fabri, who remained oblivious to the infinitude of speeds by which motion along an inclined plane must exceed the number of speeds existing in free fall, however great that number might be. Aristotle's rejection of any actual infinite in nature did not forbid incommensurable magnitudes in mathematics, and as a mathematician Fabri, had he tried counting speeds along the diagonal, and down the side, of a square, could not fail to see that no unit could measure both the quantities.

In the following theorem, proof, and commentary, Fabri seems aware of this, but unwilling to accept the mathematical solution. He evaded it by a metaphysical subterfuge:

> Theorem 61. Naturally accelerated motion is not propagated through every degree of slowness.
>
> Since there are as many of these degrees of propagation as there are

instants through which the motion endures, because new impetus is made in single instants (as shown in our *Metaphysics*), then if infinite instants were permitted, the propagation would not be through every degree of slowness that this series of degrees did not include.[107] For [a body] clearly begins to move more slowly on an inclined plane than straight down in a free medium, and in a dense medium as against a rare one (that is, more slowly in water than in air.) Therefore that degree of slowness with which it begins to move on a slightly inclined plane is not contained among those with which it moves straight down.

Fabri's commentary went on to challenge Galileo directly:

By what is said here, Galileo is rejected on two counts. First, it is in vain that he assumes infinite instants without necessity, and second, the ratio he gives is not convincing. Indeed, he calls rest "infinite slowness" from which the movable [thing] then recedes—and no doubt his motion would then proceed to be propagated through all degrees of slowness. But against him: first, rest is not in fact slowness, as it cannot have motion. Second, fast motion ensues immediately from rest as also from slow motion. Third, motion does begin, and therefore with something, so the initial motion is not infinitely removed from rest.

Le Tenneur observed that this was nothing more than a debate about terminology. The final sentence above shows that Fabri conducted it as terminologists usually proceed, since he might equally validly have concluded above "…is not somehow just a kind of rest," which was all that Galileo had meant. Fabri's metaphysical concern with words, not with physical events in the sensible world, was still more clearly shown a bit later:

There is a most certain rule that no philosopher denies: When some experiment is such that two contrary hypotheses can stand with it, neutrality can surely be deduced. Therefore Galileo cannot properly deduce his hypothesis from the experiments propounded, as I shall clearly show.[108]

---

[107] One might ask how any degree could be excluded when infinite instants were permitted with which to associate such degrees. Fabri simply refused to acknowledge that there is no difficulty for those who *do* permit infinite instants.

[108] It is not clear here whether the hypothesis meant is the law of fall, or the definition of uniform acceleration. Strictly speaking, nothing can be *deduced* from experiments, despite the wish of Newton to deduce causes from the phenomena. Much can be *discovered* from experiments, but hitting on something is not *deducing* it. Deduction is verbal, and experiments are not.

When it is said that the second distance is triple the first, assuming equal times, this is not geometrically certain and accurate... For every physical experiment must be subject to the senses, and no matter who says that, however often observed, at many places and times, it is the same..., yet since the distances are small and insensible (as are the differences) in greater, less, or equal times, spaces nearly equal to triple may be [mis]taken for triple... I ask Galileo if he or anyone else can say whether one space is triple another and, if anyone says it is off by $\frac{1}{100000}$, whether the experiment is convincing?

Galileo had answered this verbal gambit in *Two New Sciences*:

Aristotle says a hundred-pound iron ball falling one hundred *braccia* hits the ground before one of one pound has fallen one *braccio*. I say they arrive at the same time. You find on making the experiment that the larger beats the smaller by two inches... And now you would hide behind those two inches Aristotle's 99 braccia, and speaking only of my tiny error, remain silent about his enormous one.[109]

Not just fall, but the purpose of science was at stake, as well as its method of pursuit, once the times-squared law had been mathematically and experimentally established.

---

[109] *Two New Sciences*, p. 68.

Epilogue

Kepler's Problem and Galilean Units

The units devised by Galileo in the course of his discovery of the pendulum law and the law of fall were never used for any other purpose. They were not published, nor did such units occur to others, though they are singularly appropriate for the study of purely gravitational motions of any kind. Why that is so will be clear when we review the manner in which Galileo came to adopt them early in 1604; that will be done briefly at a later place in this Epilogue.

It may seem that no choice of units of length and time can be intrinsically superior for the study of gravitational motions, because all units of measure appear to be arbitrary, and any set can be related to any other. At present, two different measures of length are widely used in science—the British foot and the metric meter. However, the unit of time remains the same in both systems, this being the astronomical second.

In the British system of measurement, the acceleration due to gravity at the earth's surface is about 32 feet per second per second; in metric units, that is about 9.8 meters per second per second. There is no difficulty in converting any data from one system into the other. In practice, it might be troublesome to interpret a result in other terms if the units of both length and time were different, but in principle that does not alter the situation. Distance and time are dimensionally incommensurable, in the ordinary meanings of the words. Hence it is natural to suppose that no set of measures could have a *real* advantage over those in widespread use (or any others), but only an advantage in the recording of measures and the making of calculations. A mile, or an Ängstrom unit, would be inconvenient to use for lengths in daily use, or a time-unit of one year or one millisecond; but only because large order-of-magnitude factors would be often necessary that could easily have been avoided.

The fact that both systems use the astronomical second does not imply a time-unit any less arbitrary than our units of length. Both the foot and the meter are entirely arbitrary, being merely distances between two marks made on standard bars of some durable material, decreed by a government to serve as a legal definition. The second is no less arbitrary, for it is $\frac{1}{24}$ of $\frac{l}{3600}$ of the time of axial rotation of the earth. Those fractions originated in ancient Babylonian astrological tradition, not in any serious scientific analysis of the measurement of time.

Yet there are indeed pairs of measures, for length and time, having a real advantage over all others for the investigation of gravitational phenomena. The creation of one such pair of units was reconstructed in Chapter 5. Galileo named his units the *punto* and the *tempo*, and his determinations of them by measurements of some terrestrial gravitational phenomena were remarkably accurate. His *punto* of 0.94 mm, if my calculations are correct, was nearly $\lambda = 0.9422119204$ mm, and his *tempo* $\approx \frac{1}{92}$ astronomical second was very nearly $\tau = \frac{1}{91.88024932}$ second. Calculations using λ and τ make the acceleration due to gravitation $\frac{\pi^2}{2^3}\frac{\lambda}{\tau^2}$ $(= g\frac{\lambda}{\tau^2})$.

The "Galilean units" will be called G.U. Unlike the common foot-and-second, or meter-and-second, λ-and-τ of G.U. link time automatically with distance in all purely gravitational motions.

The square of the time in τ of fall through a distance in λ is *numerically* the same as the length in λ of the pendulum timing fall through double that distance. No such numerical identity of measures relating the times and distances of any two distinct and different gravitational motions exists in arbitrary sets of units. It is a real advantage to link time with distance in the study of planetary motions (for example) merely by adopting suitable units.

Motions of planets are purely gravitational in character, and yet not every problem that has been proposed concerning them has yet been solved. Kepler's problem—to find the rationale of the places occupied by planets—has had a very checkered history and is still pursued by those who believe that something more than chance must be involved.

In Kepler's day, Saturn was the most distant planet known to exist, and he arrived at a theory under which there could not be any more planets beyond Saturn. In 1596 he asserted all planets to move in spheres that were circumscribed around the five regular solids, centered at the sun in a certain ordering, each of those geometrical forms being used only once. In antiquity, Euclid had proved that exactly five regular solids exist, so Kepler held the number of planets to be forever limited to six. He had discovered that orbital radii fitting Mercury,

Venus, Earth, Mars, Jupiter, and Saturn were nearly proportional to the radii of the spheres circumscribing the "Platonic solids", when those were placed in a certain order with the sun at the center of all the spheres. But planets beyond Saturn have been found, so Kepler's "discovery" was no more than coincidentally almost true for the first six planets.

Two decades later, in 1619, Kepler made a genuine discovery about the planets—that the cubes of their orbital radii are as the squares of their periods of revolution. That law became the cornerstone of celestial dynamics after Newton linked it to his law of universal gravitation in 1687. The "third law"[110] threw no light upon Kepler's problem, however; rather, this weakened its status as a serious problem, because the law would hold no matter how many planets there were, or what their distances from the Sun might be.

Although Galileo appears to have made no use whatever of this planetary law, he had found another one from Kepler's first book, which was concerned primarily with the above problem.[111] By using planetary data calculated by Kepler, Galileo found the *speeds* of planets to be inversely as the square roots of their orbital radii, in 1601.[112] In 1608 or 1609 he linked this with the law of falling bodies he had discovered in 1604. From that, he conjectured that all the planets had been

---

[110] Kepler's first and second laws, in 1609, stated that planetary orbits are not circular but elliptical, and that the Sun is fixed at a focus of each ellipse.

[111] Johann Kepler, *Prodromus dissertationum cosmographicum*...(Tübingen, 1596). I translate the full title as "Prologue to cosmographical theses containing the world-structure mystery of the remarkable proportions of the celestial orbs, and the true and essential reasons for the number of skies, their sizes, and their periodic motions, demonstrated from the five regular geometric solids." There is an English translation of the second edition (of 1621) by A.M. Duncan (Alberis Books, N.Y., 1981).

[112] Though this follows from Kepler's three planetary laws, he did not deduce and state it, at any rate in the above form.

created at some place far beyond Saturn, and had reached their speeds by falling in uniform acceleration[113] toward the Sun, God having turned each to circular motion when it reached its destined speed. Newton later remarked that God would also have had to double the gravitational power of the sun at the moment each circular motion began, to hold that planet in orbit, as mentioned earlier.

Beyond Kepler's first speculation about "Platonic solids," no rule was advanced during the 17th century to link the observed distances of the planets from the sun with one another, except Galileo's notion about speeds of fall from some remote place at which all the planets could be supposed to have been created. He described it in his *Dialogue* as a way to flesh out a sketchy idea about the origin of the world found in Plato's *Timaeus.*[114]

In 1772 an amateur of science named J.D. Titius proposed an empirical rule, usually known as Bode's law[115] because the German astronomer, Johann Bode, publicized and advocated it. As it was originally expressed, Bode's law had the form:

$$\text{the orbital radius of planet } n \text{ is } 3\times 2^{(n-2)} + 4,$$

where n is the number of the planet out from the Sun. Earth, as number 3, thus had orbital radius 10; Venus had orbital radius 7, and (by a special exception) Mercury had orbital radius 4, (as if n jumped to minus infinity.) The unit implied was A.U./10, A.U. standing for the modern astronomical unit with the Sun-Earth distance = 1.

Bode's rule had the kind of simplicity desirable in a law of nature, and it appealed to numerologists because it used only the numbers up to 4—credited with mystical properties ever since Pythagoras in remote Greek antiquity. Now, when Bode's law first appeared, Saturn was still the outermost planet known. In 1781, however, Sir William Herschel discovered a seventh planet, and its distance was found to fit

---

113 Cf. my *Galileo at Work*, pp. 64–5, 154–7.

114 Of course Plato knew nothing of uniformly accelerated motion. He said that after eons of wandering in chaos, the world-bodies were put in order by the demiurge, and Galileo suggested a means of that ordering.

115 For the complete story see M.M. Nieto, *The Titius-Bode Law of Planetary Distances* (Oxford, Pergamon Press, 1972).

the next place after Saturn, $n = 8$ calculated according to Bode's law. Next, in 1801, the missing "planet" following Mars, for $n = 5$, was supplied by the asteroid Ceres, discovered by G. Piazzi during sky-searches inspired by Bode's law. Many asteroids in this region have since been found.

Bode's law thus seemed assured. Both John Couch Adams and Urbaine Leverrier assumed its truth when calculating the place of a hypothesized next planet, needed to account for perturbations of the motion of Uranus. Leverrier's calculations were so accurate that Neptune was sighted in 1846, on the very first night that the astronomer J.G. Galle searched the region in the sky Leverrier had determined for the predicted planet beyond Uranus.[116]

For a time the success of Bode's law appeared complete. But accurate measurements soon revealed that Neptune was not at the Bodeian place for $n = 9$, falling far short of that. In 1846 it happened that Neptune's position was such that distance had been unimportant in the calculations. Bode's law thus became merely a puzzle, of little interest to astronomers until 1930. Then Pluto was discovered, very nearly where Bode's law would place the next planet after Uranus. That fact restored to Bode's law its former status of astronomical curiosity. It now has too many successes, with only a single failure, for those to be merely coincidental. Something more than met the eye might lie behind Bode's law; for there is a limit to the number of lucky chances that most people regard as truly credible.

Nevertheless, there are reasons for doubting that Bode's rule can be a law of nature. Natural laws are open to discovery by any astronomer, no matter what his base of observation may be. If one planet alone, third from the Sun, moved at the crucial distance in the solar system, astronomers on Earth would have an advantage over all others in discovering the law of its structure, because the most reliable and the most convenient unit of distance for making precise measurements is the observer's own distance from the Sun.

Moreover, although the A.U. presently used is a unit firmly rooted

---

[116] Early in 1613 Neptune come into direct line with Jupiter, and Galileo noted it as a "fixed star" in his journal of observations 233 years before it was identified as a planet. See S. Drake and C. Kowal, "Galileo's sighting of Neptune," *Scientific American* December 1980.

in an actual planetary placement, the Bodeian factor of 10 is not essential in any laws of nature. The number 10 is the base of the place-notation we prefer, but 10 has not always been chosen. Ancient Babylonian astronomers, who created the science of stars, chose 60. If Bode's law were truly a law of nature, it would have eluded detection even on Earth until 1585, when decimal fractions were first introduced.

Such considerations may be useful in determining where we ought to search for whatever it is that lies behind Bode's law—for though it is hard to doubt that *something* must lie behind it, it is equally hard to imagine what *kind* of thing. Except for the place of Earth, Bode's law does not give any distance beyond two significant figures (and for Mars not even that, giving 1.6 A. U. instead of 1.5.) But out to Neptune it gives them all, including the radius of the asteroid belt, arranged in order by a function of successive integral powers of 2, without any unoccupied places. That is too much to dismiss as a parade of lucky coincidences.

To throw light on these puzzles, the link between Bode's 10 and G.U. will be shown. First, however, some of the basic data for the five planets out to Jupiter will be tabulated in m.U. (or Mercury units), putting the semimajor axis of that planet equal to 1 rather than to 0.3870987—its place in A.U. based on Earth = 1, used in astronomical tables. Distances based on that of a unique planet, Mercury—the innermost[117]—avoids fractional ratios of axes for successive planets outward from the Sun.

Writing m for Mercury and M for Mars, the principal data are presently measured as follows:

[117] It is not certain whether there is an outermost, and in the selection of units it is always advantageous to adopt a unique case if there is one.

Table 1: Planetary Data in m.U.

Semimajor axis		Period in earth days		Eccentricity of orbit		Rotations per period	
m	1	m′	87.969256	m"	0.205614	m*	1.4999
V	1.868595722	V′	224.700789	V"	0.006821	V*	0.92466
E	2.583318249	E′	365.25636	E"	0.016751	E*	365.256
M	3.936183388	M′	686.979702	M"	0.093309	M*	669.6
J	13.44051012	J′	4,332.5879	J"	0.048254	J*	10,556

Because ratios are entirely independent of the units chosen, they cannot favor observations made from a particular planet. The ratios in the first column are relevant to Kepler's problem, and to Bode's law. The first, V[/m] = 1.868595722, may look like a random assortment of digits, but it is far from being that.[118] In the logarithms called "natural" (to the base e), $V/m = 1 + \frac{2}{\ln 10} = 1.868588964$ ($= \log_{10} 10e^2$), to six significant figures. Thus the Bodeian "unit" 10 puts in its appearance at once, in the first non-trivial distance-ratio, V/m—which remains numerically the same no matter what unit of distance is adopted.

The last column is not usually regarded as containing data of any special interest, but will serve to make it clear that the Earth day, the unit in which time is measured, is really a ratio. Astronomers on other planets would use different measures of time from ours, since every planet rotates axially while it orbits the sun, and two stable measures of time suffice to establish a unit satisfactory for use in astronomical measurements.

The distance of Mercury in m.U. is, of course, trivial. But in A.U.

---

[118] So is E/m = 2.583320461 far from being a random collection of digits; 10 ln ln (eE/m) = $(E/m)^2$. It follows that its reciprocal, m/E = 0.3870987, is likewise non-random, and another independent illustration of this will be given presently.

Mercury's semimajor axis is the reciprocal of E as above, making it 0.3870987.[119] Now, suppose that instead of putting E = 1, as in A.U., or E = 10 as in Bode's law, we were to put $E = \frac{10}{2\pi}$. That will be done below. Italic initial letters will represent measures in G.U. units, while ordinary letters continue to mean measures in m.U. as in Table 1. One further assumption is seen at the head of Table 2, in order to put sidereal periods into units other than the arbitrary Earth day.

In Column 3 we have the distances of planets from the Sun in Bode's-law units. By putting $E = \frac{10}{2\pi}$ the figures in the first column had to have the same ratios to *E* as those in Column 3 have to 10, while those in Column 4 have those same ratios to 1—which is how they were measured in the first place.

Table 2: Putting $E = \frac{10}{2\pi}$

$$V' = \frac{\pi}{2}\sqrt{\frac{E}{2}} = \sqrt{gE}$$

	1 Orbital radius (in $\lambda \times 10^{14}$)		2 Sidereal period (in $\tau \times 10^{9}$)		3 Circumference of orbit	4 Radius in A.U.
m	0.616086716	*m'*	0.6989085354	*m°*	3.870987001	0.3870987
V	1.151218156	*V'*	1.785229369	*V°*	7.233317001	0.7233310
E	1.591549431	*E'*	2.901931271	*E°*	10.000000000	1
M	2.425030301	*M'*	5.457996351	*M°*	15.23691476	1.5236915
J	8.280519739	*J'*	34.42206	*J°*	52.20280400	5.2028040
S	15.18152703	*S'*	85.47300834	*S°*	95.38834756	9.5388348
U	30.52897772	*U'*	243.7929396	*U°*	191.8192243	19.1819224
N	47.83839973	*N'*	478.1858513	*N°*	300.5775303	30.0577530
P	62.76999504	*P'*	718.7368677	*P°*	394.3955106	39.4395511

[119] The autoroot of π is ρ, such that $\rho^{\rho} = \pi$; $(\pi^2 + \rho^2)/\rho^2 = 10m/E$ to 6 significant figures, at 3.870983.

Orbits are elliptical, as Kepler was the first to discover. When we take the orbits to be circles, with the semimajor axes as their radii, $E = \frac{10}{2\pi}$ simply transforms the Bodeian planetary distances into orbital circumferences. That is a mathematical consequence of the reciprocal relation between m.U. and A.U., not a "cosmological mystery" of the kind Kepler attempted to penetrate in his first published book. But it is not a merely mathematical consequence that the ratio $\frac{E}{m}$ and its reciprocal $\frac{m}{E}$ are related to the ratio $\frac{10}{2\pi}$ astronomically, and hence gravitationally, and thereby physically. Mathematics simply makes it easy to recognize astronomical relationships that would otherwise go unnoticed.

It alters the nature of puzzles about Bode's law when we know that the unit of solar distance it implied for the planets has a real meaning, though not the meaning that Bode's law derived from it. The number 10 which Bode's law gave to the Sun-Earth distance was a step in the right direction. The next move is to bring in π; or rather, to put π, 2, and $\sqrt{2}$ into our *units* and automatically link our measures of distances with measures of times through gravitational phenomena. But that is done by Galileo's constant.[120]

The symbol 10 is involved in point-shifts of one place either way,[121] usually associated with a difference in order of magnitude. When working with astronomical magnitudes it is often necessary to move the point many places, symbolized by the use of $10^{\pm n}$. If we used binary arithmetic, that same *symbol* would be used to denote order of magnitude, but it would then *mean* $2^{\pm n}$, which lies at the very heart of Bode's law.

Next, neglecting position of the point, in decimal notation the *digits* remain the same in $\sqrt{2}$, $\sqrt{200}$, etc. or in $\sqrt{20}$, $\sqrt{2000}$ etc.; but the digits in

---

[120] Galileo's constant is π over 2 times the square root of 2; its square is g, the universal acceleration due to gravitation in G.U.

[121] In any place-system of numeration, that operation uses the symbol 10, but that symbol does not mean the number ten except in our decimal system. In binary notation, for example, 10 means the number two, and that number appears in nearly all natural laws.

the two sets differ by the factor $\sqrt{10}$, or $10^{1/2}$. Since in gravitational phenomena, all distances are as the square roots of certain times, we would expect units automatically linking every distance with some time, when expressed in binary notation, to mingle together two things we regard as different in kind—the order of magnitude factor, 2, and the factor which can preserve digital integrity in sets of square roots after one-place point-shifts, which in binary notation is the square root of two. Kepler's third law and Bode's law may be related in a way concealed from us by our habit of taking the *symbol* 10 always to mean the number ten. The exponent $3/2$ in Kepler's law relating distance-ratios of planets to their sidereal period, when applied in G.U. is $\ln M/m$.

There is nothing remarkable about Galileo's having used 2 and $\sqrt{2}$ in his discovery of the pendulum law, when we now look back at Chapter 5. He used doubling of lengths because that is easy to do with precision, and it did not lead immediately into impracticable lengths in experimental measurements. He used mean proportionals mathematically equivalent to $\sqrt{2}$, because the rigorous Euclidean theory of ratio and proportion for mathematically continuous magnitudes—not just for numbers—suggested that to him. In retrospect, it ceases to be puzzling that Galileo arrived at the unique constant of gravitation that is the same everywhere in the solar system, though that seemed the most puzzling fact revealed by his working papers when it was first noticed a few years ago.

What is most noteworthy in Table 2 is that the distances in Column 1, if we neglect orders of magnitude shown at the top, are almost exactly in G.U., with $m \approx \frac{g}{2} = \left(\frac{\pi}{4}\right)^2 = 0.6168502751...$ out to infinity. It was for that reason that V' was assigned the value $\sqrt{gE}$, relating G.U. directly with m.U. To show how those two sets of units may be physically related requires a digression here into pendulums and fall.

From the modern equations for distance fallen from rest, f, and for time of pendulum p to the vertical through a small arc,

$$f = g\frac{t^2}{2} \text{ and } t = \frac{\pi}{2} \times \sqrt{p/g}$$

it is easily deduced that when any fall from rest is timed by a pendulum-swing to the vertical, distance fallen is $\frac{\pi^2}{8} \times$ the pendulum length;

and that when any length is specified, that pendulum takes $\frac{\pi}{2\sqrt{2}}$ times as long to reach the vertical as a body takes to fall from the height equal to the length of the pendulum. It was, in fact, the latter measurements that gave to Galileo the pendulum law, a day or so before he discovered the law of fall from it. It is the ratio $\frac{\pi}{2\sqrt{2}} = \sqrt{g} = 1.11072$ that I called "Galileo's constant," of which the square, g, is the gravitational acceleration in G.U.

Galileo's *punto* and *tempo* entailed acceleration not quite g (= 1.2337), but close to that, as 1.228 *punti/tempo*^{2}. He could not apply his units to planetary data, because at his time there were no measurements of celestial distances in conventional units such as miles. Only their *ratios* were measurable. That is not so today. Mercury, 57.9 million km from the Sun, has the "orbital radius" 0.616×10^{14} *punti*. That is almost exactly the measure in G.U. shown for Mercury in λ, in Table 2.[122]

Next, turning our attention to the unit of time, the sidereal period of Mercury is 87.969256 Earth days, or 0.69834×10^9 *tempi*, which is nearly m′ in the same table. It is accordingly clear why the name "Galilean units" is appropriate for measures of length and time that make g exactly $g = \frac{\pi^2}{2^3}$, and also why G.U. are of particular use for relating orbital radii and sidereal periods.

It is a remarkable property of G.U. that in its measures, the square of any *time* of swing to the vertical through a small arc by a pendulum of length $\pi\lambda$ is the *length* of pendulum timing fall from rest through distance 2 $\pi\lambda$. Hence when purely gravitational phenomena (such as fall and the swing of a pendulum) are measured in G.U., lengths and times are no longer *incommensurable* in terms of physical science. Also, in calculations performed when data are in G.U., the time of fall at acceleration g through orbital radius r is $\sqrt{r}$, the square root of that same radius when measured in m.U.—neglecting orders of magnitude. If we take those also into account, a point-shift of two places is required in order to have 10^7 (the square root of 10^{14}) apply to the peri-

---

[122] Unfortunately the mean distance in kilometers is reported only to 3 significant figures, so though we can put the *punto* at 0.94 mm, we need more information to make calculations beyond 3 places.

odic times.

When we calculate the time for half-swing of the pendulum as long as is the semimajor axis of Earth in G.U., we find this to be almost the same as $\frac{V'}{10^2}$ as shown in Table 2. Taking Venus' period in Earth days, $V' = 224.700789$, and dividing by 24 (hours per day) and 3,600 (seconds per hour), we find τ implied to be almost exactly Galileo's *tempo* $= \frac{1}{92}$ of a second of time.

It suffices here to have indicated the importance of π in the sky, and the fundamental roles of squaring and point-shifting in planetary relations of the kind I call non-Keplerian.[123]

There exists a convergent series of operations, by which any number greater than 1 will become E/m; it is the series:

$$\ln, +1, \ln, 10, \sqrt{\ }, \text{repeated.}$$

This converges to 2.583322768, which will remain unchanged. It is as if $E/m = 2.583320481$ were a gravitational maintenance of this relation, rendering stable the orbits of the first and third planets. Neglecting relative orders of magnitude, the square of the ratio E/m is nearly G, the Newtonian constant of gravitation.

The *kind* of relation sought by Kepler geometrically, and in Bode's law arithmetically, in which each solar distance leads to the next, appears to exist with respect to the first four planets only. It is $y = (\ln \ln 10x^2)^4$, in which x being V/m, $y = E/m$; and x being E/V, $e^x = M/m$. Because $V/m = 1 + \frac{2}{\ln 10}$ to 6 significant figures, while $\frac{10m}{E} = \frac{\pi^2 + \rho^2}{\rho^2}$ to the same degree of accuracy, ρ (which is 1.854105969) being the autoroot of π, such that $\rho^\rho = \pi$, the first four semimajor axes in A.U. can be obtained from π, the number 2, exponential operations, squaring, and point-shifting.

Kepler's problem may have been a valid one after all, and the Galilean units may play an essential role in its solution. If so, the final chapter in the history of fall has not yet been written by scientists.

---

[123] For more information on such relations see my *Galileo: Pioneer Scientist*, University of Toronto Press (in press).

Bibliography of Works Cited

Albert of Saxony — *Subtilissimæ quæstiones super octos libros Physicorum*, Venice, 1504

Aristotle — *De cælo*

Physica

Metaphysica

Aristotle (?) — *Quæstiones mechanicorum*

Baliani, G. B. — *De motu gravium solidorum*, Genoa, 1638, 2d ed. Genoa, 1646

Benedetti, G. B. — *Demonstratio proportionum motuum*, Venice 1554

Diversarum speculationum, Torino, 1585

Borro, G. — *De motu gravium et levium*, Florence, 1578

Bradwardine, T. — *Tractatus de proportionibus velocitatem in motibus [1328]* ed. B. Politi, Venice, 1505

Buridan, J. — *Quæstione super octo physicorum libros Aristotelis*, ed. J. Dullaert, Paris, 1509

Cardano, G. — *De subilitate*, Nuremberg, 1550

Cavalieri, B. — *Lo Specchio Ustorio*, Bologna, 1632

Clagett, M. — *The Science of Mechanics in the Middle Ages*, Madison, 1959

Crosby, H. L. — *Thomas of Bradwardine, his Tractatus de Proportionibus*, Madison, 1955

Drabkin, I. E. — *A Source Book in Greek Science* (with M.R. Cohen), New York, 1948

Galileo On Motion and On Mechanics (with S. Drake), Madison, 1960

Drake, S. — *Mechanics in 16th-Century Italy* (with I.E. Drabkin), Madison, 1969

Galileo at Work, Chicago, 1978

Bradwardine's function, mediate denomination, and multiple continua, *Physis* 12: 1, 1970

Galileo's experimental confirmation of horizontal inertia, *Isis* 64 (1973), pp. 291 ff.

The Role of Music in Galileo's Experiments, *Scientific American*, June 1973

Euclid — *Elements*, tr. B. Zamberti, Venice, 1505, tr. N. Tartaglia, Venice, 1543, tr. C. Clavius, Rome, 1578

Fabri, H. — *Tractatus physicus de motu locali*, Lyons, 1646

Fermat, P. — *Oeuvres*, ed. Tannery & Henry, Paris, 1891–1912

Galileo — *Dialogue Concerning the Two Chief World Systems*, tr. S. Drake, Berkeley, 1953, 1967

Opere, ed. A. Favaro, Florence, 1890–1909

Two New Sciences, tr. S. Drake, Madison, 1974, 2d ed. Toronto, 1989

Heytesbury, W. — *Regule solvendi sophismata* [Oxford, 1335]

Hipparchus — Περι τον δια βαρυτητα κατο φερομενων *[*as cited by Simplicius]

Kepler, J. — *Prodromus Dissertationum Cosmigraphicarum continens Mysterium Cosmographicum de admirabili Proportione Orbium Coelestium ... Tübingen, 1596*, 2d ed. Frankfort, 1621; tr. A.M. Duncan, New York, 1981

Le Tenneur, J-A — *De motu accelerato*, Paris, 1649

Mach, E. — *Die Mechanik in...Entwickelung*, Leipzig, 1889

Marchia, F. — *Reportacio* on Book 4 of Peter Lombard's "Sentences"

Monte, G. del — *Mechanicorum liber*, Pesaro, 1577

Nieto, M. M. — *The Titius-Bode Law of Planetary Distances*, Oxford, 1972

Oresme, N. — *De proportionibus proportionum*, ed. E. Grant, Madison, 1971

Pereira, B. — *De communibus omnium rerum*, Venice, 1586

Settle, T. — An experiment in the History of Science, *Science* 133 (1961), pp. 19–23

	Galileo and early experimentation [in *Springs of Scientific Creativity*, ed. Aris, Davis, & Steuwer, Minneapolis, 1983]
Simplicius	*In Aristotelis de Cælo commentaria*
Stevin, S.	*Works*, ed. E. Dijksterhuis et al, Amsterdam, 1955–66 (Dutch and English)
Tartaglia, N.	*La Nova Scientia*, Venice, 1537
	Quesiti, et invwentioni diverse, Venice, 1546
	Regola generale da sulevare...ogni affondata nave, Venice, 1551
	Jordani opusculum de ponderositate, Venice, 1565
Varro, M.	*De motu tractatus*, Geneva, 1584
Wallace, W. A.	The concept of motion in the sixteenth century, *Proceedings of the American Catholic Philosophical Association*, Washington, 1967
	The enigma of Domingo de Soto, *Isis* 59: 4 (1968)
Wilson, C.	*William Heytesbury: Medieval Logic and the Rise of Mathematical Physics*, Madison, 1956

Index

A

acceleration, 3, 7, 9, 12, 17-19, 21, 26, 30, 32, 62-63, 70, 73, 77, 81-82, 84, 91
  as loss of force, 30, 62-63
  odd-number rule of, 1, 37, 68, 75
Adams, J.C., 85
Adelard of Bath, 23
Albert of Saxony, 12, 19, 26, 68, 71, 75-76
  See also rule of
Alexander of Aphrodisias, 30
algebra, 15
antiperistasis, 52
Arabs, 12, 18, 23-24
Archimedes, 5-6, 24, 27, 33, 70, 91
  principle of, 27-28, 52
Aristotle, 2, 4, 6-16, 20, 23, 25, 28, 36, 52-53, 55, 77, 79
  De cælo, 7-8
  Metaphysics, 11
  Physics, 8-9, 11, 13, 52
  physics of, 5-6, 8, 10-12, 25-26, 29, 52, 91
  principles of, 10, 20, 25
  Questions of Mechanics, 53-54
artillery, 25, 53, 60

B

Baliani, G.B., 5, 69-70, 75
Beeckman, Isaac, 71-72
Benedetti, G.B., 27-30, 32
Bessarion, Cardinal, 24
Bode's law, 84-90, 92
Borro, Girolamo, 32
Bradwardine
  Bradwardine's function, 15
Bradwardine, Thomas, 13-16
  Bradwardine's function, 15-16
Buridan, Jean, 2, 12, 18-19, 21, 76

C

Campanus of Novara
  See Euclid
Castelli, Benedetto, 51, 69
Cavalieri, Bonaventura, 51, 65
Cazré, Pierre, 70, 73-75
Clagett, Marshall, 5
Clavius, Christopher, 25
continuity, 23-24, 53, 71-72, 76-77
Copernican system, 64

D

decimal fractions, 29, 46, 86
Descartes, René, 3, 71-73
Doni, G.B., 70
dynamics, 3, 83

E

Ecclesiastes, 25
Einstein, Albert, 12
Euclid, 23-25, 27, 29, 43, 48, 72, 75, 82, 90
  Campanus version, 23, 25
  Elements, 23-24
  Elements, Book I, 75
  Elements, Book V, 42, 77
  Tartaglia version, 25
  Zamberti version, 24
Eudoxus of Cnidos, 23
experiment, 28, 32-33, 35, 41, 43, 51, 56, 58, 60-61, 68-71, 73-74, 78-79, 90

F

Fabri, Honoré, 3, 70-71, 76-78
fall from the moon, 67
  See also law of fall
fall, actual, 2-3, 16, 18, 20-21, 28, 44, 49, 69-70
Fermat, Pierre, 71, 75
Florio, John, 58
force, nature and, 10, 33, 52, 61
forced motion, 8-10, 13, 15

G

Galilei, Galileo, 1-5, 14, 24-25, 29-33, 35-44, 46-49, 51-56, 58-65, 67-79, 81-84, 89-92
  constant of, 48, 89-91
  Dialogue, 1, 51, 64-65, 67-68, 70, 84
  Dialogue on Motion, 29-30, 35, 51, 70

La bilancetta, 29
units of, 1, 3, 31, 39, 41, 46, 81-83, 85, 87, 89-92
working papers of, 43
Galle, J.G., 85
Gassendi, Pierre, 70, 73-75
gravitation
acceleration of, 91
constant of, 90, 92
energy of, 61-63
terrestrial, 82
universal, 65, 83

H

heaviness, 3, 8-9, 19, 21
Herschel, Sir William, 84
Heytesbury, William, 17-18, 21
See rule of
Hipparchus, 7-12, 18, 30-31
Huygens, Christiaan, 3-4

I

impetus, 12, 18-19, 21, 26, 30, 33, 71, 75-77
and discontinuity, 77
impressed force, 3, 8, 18, 30, 54, 61
inclines, 36, 49, 55-56, 59-61, 64, 77-78
indeterminacy, 77
indivisibles, 72
inertia, 56
infinitude, 20, 76-77

J

Jordanus Nemorarius, 29

K

Kepler, Johann, 64, 82-84, 89-90, 92
Kepler's problem, 3, 81-83, 85, 87, 89, 91-92
third planetary law of, 83
kinematics, 10, 12, 26, 72, 75-76
kinesis, 6, 11, 17

L

law of fall, 1-5, 7, 26, 35, 37, 39, 41-47, 49, 51-52, 54, 56, 64, 67-68, 70-71, 73-75, 81, 83, 91
Le Tenneur, Jacques-Alexandre, 71, 76, 78
Leaning Tower of Pisa, 31
Leverrier, Urbaine, 85
locomotion, 6, 11, 17

M

Mach, Ernst, 73
magnitudes, ratios of, 23
Marchia, Francesco di, 12-13, 18, 30
mathematical physics, 5, 14, 16, 64, 77, 91
and quantum theory, 77
and symbolism, 15
matter, 10-11
measures, 58, 81-82, 89
accuracy of, 37, 91
and procedures, 81, 87
medium, 7, 27, 29, 31, 52, 78
Mersenne, Marin, 69-73, 76
Merton College, 13, 16-17
Merton rule, 17-18
Metaphysics, 10
middle-degree postulate, 17
Monte, Guidobaldo del, 35, 49, 53
motion
cause of, 3, 8, 18, 30
circular and straight, 25, 65, 84
composition of, 53, 58
conservation of, 61
continuity, 76
instantaneous, 55, 73-74, 76
natural, 11, 19-20, 30, 37, 52, 61, 69
resistance to, 14-16, 18-19
speeds of, 13-14
tendencies to, 8-10, 18, 26, 52, 58, 61
uniform, 9, 17-18, 21, 65, 68
uniformly difform, 17, 19, 21, 26
See uniformly difform motion
motive force, 9

N

natural philosophy, 4-5, 12, 14, 18, 26, 28, 58, 65
net forces, 15
Newton, Isaac, 2-4, 12, 32, 65, 69, 83-84
numbers
ratios of, 24-25
square, 21, 49
triangular, 21
numerologists, 84

O

observations, 10, 28, 54, 87
one-to-one correspondence, 75-76

P

parabolic trajectory, 49, 51, 56, 58
pendulums, 5, 35, 40-48, 70, 82, 90-91
law of, 1, 3, 5, 43-44, 46, 56, 70, 81, 90-91

percussion, 54
Pereira, Benedict, 30
Philoponus, 9-10, 12, 30
Piazzi, G., 85
Pisan de Motu, 30, 53
Plato, 11, 84
  Platonic cosmogony, 64
  Platonic solids, 83-84
  Platonism, 14
  Timaeus, 84
power, motive, 11, 13-14
procedures, 58-59
projectiles, 8-9, 49, 52, 61
projection, 8-9, 25, 30, 54, 56, 58-61, 63
  oblique, 51, 61, 63
proportionality, 2, 7-8, 14-15, 39, 42, 65, 72
  definition of, 2
  medieval theory of, 14, 23
  of speeds to time, 2
  of speeds to times, 7, 54, 68
  of speeds to weights, 7
  roots of distances, 56, 83, 91
  to distances, 7, 73, 76
  to squares of times, 7
Pythagoreans, 11, 24

R

ratios, 14, 23, 25, 39, 46, 48-49, 64-65, 86-88, 90-91
  compound, 48-49
Ricci, Ostilio, 25
Roberval, G.P. de, 72

S

Sarpi, Paolo, 74
Scheiner, Christopher, 67-68
Settle, Thomas, 32
ships, 27
Simplicius, 8, 12, 30
Soto, Domingo de, 26
specific gravities, 27-28
speeds
  Aristotle on, 3, 7, 19
  equal, 9, 12, 28-29
  horizontal, 56
  in water, 27-29
  increase of, 20, 33, 37
  instantaneous, 54
  momenta of, 55, 74
  of planets, 2, 64, 83-84
Stevin, Simon, 28-29, 32

T

Taisnier, Jean, 28
Tartaglia, Niccolò, 24-27, 29, 32, 53
Theon of Alexandria, 25
time-intervals, 21
Titius, J.D., 84

U

uniform speed, 21, 56, 76-77
universities, 23-25
University of Padua, 44

V

Varro, Michael, 33, 68
vellum, 60
vis derelicta, 12-13, 19
Viviani, Vincenzio, 31
void space, 52

W

Wallace, William, 26
weight, 7, 10, 13-14, 31-32, 39-41, 43, 54, 60

Z

Zamberti, Bartolomeo, 24
Zeno's paradoxes, 55